NUMINOUS SEDITIONS

NUMINOUS SEDITIONS

INTERIORITY AND CLIMATE CHANGE

Tim Lilburn

UNIVERSITY of ALBERTA PRESS

Published by

University of Alberta Press
1–16 Rutherford Library South
11204 89 Avenue NW
Edmonton, Alberta, Canada T6G 2J4
amiskwacîwâskahikan | Treaty 6 |
Métis Territory
uap.ualberta.ca | uapress@ualberta.ca

Copyright © 2023 Tim Lilburn

LIBRARY AND ARCHIVES CANADA
CATALOGUING IN PUBLICATION

Title: Numinous seditions : interiority and
 climate change / Tim Lilburn.
Names: Lilburn, Tim, 1950– author.
Description: Includes bibliographical references
 and index.
Identifiers: Canadiana (print) 20230227090 |
 Canadiana (ebook) 20230227139 |
 ISBN 9781772127102 (softcover) |
 ISBN 9781772127256 (PDF) |
 ISBN 9781772127249 (EPUB)
Subjects: LCSH: Environmentalism—Philosophy. |
 LCSH: Poetry. | LCSH: Ecocriticism.
Classification: LCC PS8573.I427 N86 2023 |
 DDC C814/.6—dc23

First edition, first printing, 2023.
First printed and bound in Canada by Houghton
Boston Printers, Saskatoon, Saskatchewan.
Copyediting by Tania Therien.
Proofreading by Jenn Harris.
Indexing by Stephen Ullstrom.

All rights reserved. No part of this publication may be reproduced, stored in a retrieval system, or transmitted in any form or by any means (electronic, mechanical, photocopying, recording, or otherwise) without prior written consent. Contact University of Alberta Press for further details.

University of Alberta Press supports copyright. Copyright fuels creativity, encourages diverse voices, promotes free speech, and creates a vibrant culture. Thank you for buying an authorized edition of this book and for complying with the copyright laws by not reproducing, scanning, or distributing any part of it in any form without permission. You are supporting writers and allowing University of Alberta Press to continue to publish books for every reader.

This book has been published with the help of a grant from the Federation for the Humanities and Social Sciences, through the Awards to Scholarly Publications Program, using funds provided by the Social Sciences and Humanities Research Council of Canada.

University of Alberta Press gratefully acknowledges the support received for its publishing program from the Government of Canada, the Canada Council for the Arts, and the Government of Alberta through the Alberta Media Fund.

CONTENTS

Preface VII

1 | New Sadness 1
2 | Interiority and Climate Change 11
3 | Contemplative Practices, Contemplative Pedagogies 25
4 | Hoping for Something to Appear 49
   The Poetry of Don Domanski
5 | Poetry's Practice of Philosophy 55
   Anne Szumigalski
6 | Reading William Chittick Reading Ibn 'Arabi 65
7 | Happy Incompetencies, the Self's Other Routes 77
8 | Poverty and the Doom of Acedia 91
9 | Ontological Loneliness and the Balm of Metaphor 103
10 | Two Readings on Snow, Two Readings on Sorrow 113
   Reading Gillian Rose
11 | In the Time of Extreme Heat, In the Time of the Discovery of Unmarked Graves at the Sites of Residential Schools 131
12 | Numinous Seditions 147

Dream Coda 159
Glossary 161
Reading 163
Index 169

PREFACE

LAST JUNE I WATCHED a patch of mustard I had planted in the University of Victoria community garden, and watered daily, simply vanish over a few days of very high heat. Looking at that rectangle of dirt become cleared of living things, I saw with a jolt of deep emotional clarity where we are—in the midst of a dramatically and irreversibly changed circumstance. Part of the blow of the coastal heat dome in 2021 was the realization that this extreme heat, with its plant, animal and human deaths, was only the beginning, that this deadening heat would repeat itself and worsen, and that it would be only a part of a cluster of disasters including great fires, biblical flooding, which came in November, and season-long droughts.

Few people went out in those uncannily hot days, and if they did, they stayed close to shade. Big boughs of Garry oaks exploded because of the temperatures and came crashing down. One missed me by a couple of yards as I sprinted from the loud crack. The leaves on all of the oaks on the west side of the mountain behind my house turned brown in the pounding light. As the heat continued in the high 40s, fires broke out in Lytton and Princeton and other places in British Columbia; whole towns burnt to the ground. People at the community garden covered their plants with cotton sheets or they set up shade shelters of roofing shakes or staked old frisbees, but nothing worked and much of the planting died off, beans, kale, spinach and peas, or grew stunted.

We are entering a new time, a changed climatic and cultural surround, and we don't, at the moment, know how to move in it. And we don't know where to go to learn how to be under the new conditions of an altered climate, most of us locked in a stasis of stunned amazement. Damascius, a late Platonist, during the crisis caused by the emperor Justinian's suppression of the Academy of Plato in Athens in 529, saw, with alarm, that a sustaining world, a feeding culture, now stood on the razor's edge. We have a similar sense of precariousness. A way of life and the thought that sustains it, the thought of the Enlightenment, the sense of a teleological force animating history, as conceived by Augustine, Hegel, Mill and Marx, favouring economic and cultural advance, is vanishing, and we are entering an orphaned state. Damascius's response to his pancultural dissolution was to try to gather as many parts of the disappearing world as possible and set these against the cataclysm. This preservation of thought-palaces, rituals, a form of common sense, particular forms of listening and valuing, which he called *patria* and recorded in his book *The Philosophical History*, was intended as a way to withstand complete loss, but soon became simply a way to mourn. There is no antinomy between a practice of sadness and an engaged politics at certain times of severe crisis, as Ashlee Cunsolo (169–89) and Gillian Rose, whom we will meet later, make clear.

Numinous Seditions attempts something similar to Damascius's collecting of cultural fragments, but the thought-world it wants to preserve must be first retrieved, drawn out of the same tradition Damascius wanted to protect, the tradition that saw deep thought, philosophical thought, as a form of life, a deposit of therapeutic wisdom.

✶ Big tech will save us. Colonization of other planets, some say, will save us. The abolition of the fossil fuel industry will snatch us away from the edge. There are elements in each of these solutions, aside from the new planet idea, that seem plausible. But no alterative can work on its own. I am caught by another focus—how may our inner lives be shaped so that we can go on and, in a sense, thrive in these disarming times? What are our ways of interior perduring?

There is no comprehensive program for such a novel inner formation in the era of loss we now enter; we don't know enough to hatch such a thing. We need the suppleness of discrete muse-goaded, agent-intellect-propelled ruminations, to probe a way ahead. Passage into a contemplative life in the time of global warming amounts to a series of numinous seditions, audacious renovations, expansions, in ways of seeing, thinking and conviviality that also have the effect of stretching the way people see the places they are, their ontologies. These shifts, as we take them on as interior exercises, will deposit us outside of the modern world, that is, the capitalist, consumerist world of the twentieth and twenty-first centuries, the world of much of canonical Western philosophical thought over the last four hundred years, the world of sanctioned devotion from Scholasticism on. I hope that the essays that follow will help pick out parts of this interior, revolutionary way and assist people on it. I am not a scholar, as I have admitted in previous books, but, as I like to think, a panic-struck citizen who happens to have a library card. The essays contain some of the elements I have found of a form of life that is able to move within the sorrow, loss and disarrangement of what is upon us. Who can say how we will live in the years to come? Nevertheless, we at least need now a refugium for attentiveness and plunging thought, an instance of *statio*; I propose one made up of certain forms of philosophy, poetry and mystical theology and the confluence these means of exploration enjoy.

Most of the essays in *Numinous Seditions* grow out of conversations at conferences, in coffee shops, classrooms and on the trail. I wish to thank Helen Marzolf, Jan Zwicky and Philip Kevin Paul for being my primary interlocutors over the last few years when many of these essays were written. I thank Erica Grimm at Trinity Western University for involving me in discussions with climate change scientists at a memorable gathering in 2015 and Sheri Benning for collaborating with me in the design and offering of the first "Writing into Climate Change" workshop at St. Peter's College in 2018. I also thank Louise Halfe and Peter Butt for many discussions in the lodge and on the phone and for introducing me to the late Elder Joe Cardinal, to whom I am indebted for his hospitality and his teaching. I am still learning from what Joe said to me. I thank Andi Lloyd and Jonathan Miller-Lane for making possible the three-part lecture series "Contemplative Practices, Contemplative Pedagogies" I gave at Middlebury College through the fall term in 2018 that helped clarify many of the ideas in this book. I am also in debt to the poet Xi Chuan for inviting me to Beijing Normal University in 2015 to talk about poetry, epistemology and the self and to G.C. Waldrep for his poetry, his many letters on poetry and the spiritual life and his friendship. I thank Kevin Paul for his friendship and for leading me into the beauty of SENĆOŦEN and for his notion of "authentic curiosity." I thank SN, AKE for its instruction. And the hermit James Gray, OSB, for his instruction. Finally, I thank Peter O'Leary for his generous support of my work and his readings of apocalypse and Roberto Harrison for his vision of the Tecumseh Republic.

Hypomnemata, personal notes, interior colloquy, exhortations to the self, is a style of writing from antiquity. Aspects of this writing form are present in *Numinous Seditions*, but the book seeks its listeners and interlocutors. In writing of this type, contemplative writing, there are inevitable repetitions as the same interior blockages are approached from a variety of angles. I recommend a reading of small bites in the work that follows.

NEW SADNESS

I OFTEN FIND MYSELF ON THE OUTSIDE, in a place most are not. This location can be a matter of embarrassment and bewilderment, "confused, confused," as the *Tao Te Ching* says. Elizabeth Bishop is reported to have attempted a lifelong imitation of a "normal woman." I understand a version of the aspiration.

I have become a student of a language spoken by fewer than twenty-five people; conceivably, soon it will be spoken by no one. When I told friends in China I was doing this, they were astounded. "A magical language!" exclaimed Beijing poet Hu Zhu Dong. Perhaps. I believe SENĆOŦEN, first language of the Saanich Peninsula, to be a chthonic tongue and indeed in possession of unusual powers.

At the same time, I am trying to exist in an intellectual culture that more or less has vanished completely, buried for several centuries beneath traditional philosophy, systematic theology, the refusal of ecstatic metaphysics, questionable Plato scholarship, the hegemony of a narrowly construed rationality—the West's contemplative tradition, which is really the mystical Pythagorean philosophy of Plato and its remarkable branching into monastic Christianity, esoteric Judaism and medieval, Neoplatonic Sufism. I attempt to live in the world this culture can create and sustain for aesthetic and what I believe to be urgent political reasons. These places of standing—SENĆOŦEN, Bernard of Clairvaux et al., the Zohar, Ibn 'Arabi—encourage an ascesis of looking that could found a life, which, I recklessly suppose, approaches autochthonicity, working that unlikely alchemy in which one arises, dumbfounded, from a comprehending encounter with singularity, the individualities of things, pierced and altered.

✽ Climate change is studied from a variety of vantage points. Marine biologists, specialists in glacier deterioration, climatologists, economists all give it attention. Ample studies show a situation exists that is unprecedented in human history; this condition is multifaceted, including ocean acidification, polar ice melt, extreme weather events, unprecedented fires in the Russian and Canadian boreal forests, great fires in Australia, species collapse, coastal flooding, the displacement of a significant part of the earth's population and much more. An alteration of the earth the equivalent of an asteroid strike spread over a few generations and projecting into the foreseeable future is underway. It is not certain we know how to stand within this. The interiority appropriate to a catastrophe of such magnitude does not lie within the bounds of our current imagination.

To be sure, climate change is a scientific, economic, political and technological cluster of problems, but the psychagogic or interior aspects of it must also be considered. The biospheric problems are related, but not reducible, to colonialism, and autochthonicity is in part an attempt to treat this mass spiritual malady. But "living in the world as if it were home," even if this extreme human feat were remotely possible, cannot alone meet this new pan-ontological phenomenon. It could not manage alone to bear the full load of this new sorrow. The shape of this putative cure or balm, no less than the weight of the new sorrow, makes autochthonous practice alone not sufficient to meet the new conditions.

✽ I attended a conference on art and the environment in October 2015, at Trinity Western University, which lies in the lower reaches of the Fraser River, one of the world's greatest salmon rivers, in the heart of traditional Stó:lō territory. At the conference, various scientists spoke on current and predictable shell formation in crustaceans; plankton health; ocean warming and acidification; and on glacier melt in Greenland to an audience of other scientists, poets and visual artists. The talks were troubling, but I found it even more disconcerting that the scientists did not leave immediately after their presentations but stayed until the last cup of coffee and dry sandwich were consumed. It was a little like receiving a Saturday morning phone call from your doctor wanting to talk about your test results. I had assumed that the scientists would suppose that they were presenting their work chiefly to the other scientists attending the conference and that they would decamp to the nearest bar or coffee shop once they had done this. But it was evident as the day progressed that they were anxious to recruit the artists present to alert the public about their distressing, in some cases horrific, findings. They felt they were not able to do this themselves without undermining the appearance of being absolutely without bias in their fields. They were clearly alarmed by what they were discovering and wanted as many people as possible to learn about the dire realities that had surfaced in their work.

The reaction of some of the artists also was unexpected. They were flattered by the attention coming from an unexpected place, of course, and, as thinking citizens, they were quite willing to support the task of environmental activism; indeed most were already involved in such activism in their communities. But some demurred before the overture. They were cautious, I suspect, about bringing too much thematic and political intentionality into their art practice, believing it could jeopardize an element of serious play central to their discipline. This reduction, they sensed, would have not only significant aesthetic effects on their work, but political ones as well.

Three ludic elements are in dynamic operation in creative enterprise. The first is the work of metaphor formation, the linking of unlike things in a performance, which, though it may be beautiful, is more than merely decorative. Without architectonic intention, metaphor fashioning, following a non-reductive homological intuition, forms surprisingly heterogeneous wholes. These appear as by-blows of the attempt to elucidate emotional states and the apperception of *haecceities*.

A second form of serious play, a dionysian thinking, within art practice involves absorption. Lyric art-making commands a stance of persistent, charged, purposeless attention, spent in the contemplation of striking individualities, especially, most tellingly, the least obvious ones, the leaf at the edge of the frame, the deer with the injured foot, the blue of a dress. A third element in the shaping of poems, the completion of canvases, is the enactment of a floaty, drifting, non-compulsive, multi-channel mimesis in an effort to receive and replicate the real as it appears in the charged, affect-rich atmosphere of contemplative attention. These moments of crucial play are epistemological stances, carrying tacit political initiatives.

Latent within metaphoricity, transfixity and mimesis is a way of life—the heterodox epistemological practice of an omnivorous attention, the forming of heterogeneous gestalts as the unintended results of effort to draw human emotion into arrested looking—and a politics. These political and daemonic powers could be diminished or imperiled if too heavy a charge of will-directed shaping entered the discipline, and the art would have less striking maieutic effect as a result. Serving as an amplification device to climate change science could involve a net political loss in the reduction of *catanyxic* power in metaphor making and contemplative noting. This is not to say that science-inspired art, or political art, is impossible. I think with admiration of the poems of A.R. Ammons, Clayton Eshleman, Adrienne Rich, and Christopher Dewdney, as well as the photography of Edward Burtynsky and the land art of Peter von Tiesenhausen and Marlene Creates. But in none of these is the serious looseness in mimetic tracking compromised by trading significant parts of it for intent; but it seemed to me the scientists were asking something sharper from us, and this would produce a reduced political charge in the art.

✳ An ordo of the formative practices endemic to art-making, while it may promote a deeper dwelling in things, like the ascetical activities at the heart of the quest for autochthonicity, still does not give us, I think, the interior posture needed to carry the weight of the new sadness full climate change will bring upon us. Even these liberated cognitive architectures lack that tensile strength or lack the appropriate constitution or morphology. Something more is needed. Do we have any precedents? No, but there are distantly analogical situations that might serve as approximations to our dawning condition, though they arise as interior behaviours in reaction to quite other disasters.

The nature of the sadness that is and will be experienced in the face of the effects of global warming, which people at the conference listening to dire scientific reports received in the form of a foretaste, instantly struck me as unlike anything in memory or imagination. It occupies an entirely new category. Though it may contain aspects of malaises we know quite well, like regret, nostalgia, *penthos*, depression and despair, there was an unnamed something else; it seems as a whole to be other than conditions we are familiar with, other even than these in novel arrangement, with an unidentified intensifier, so that it is an entirely new sorrow. Because the cataclysm is unprecedented, understanding of response to it remains undiscovered. However, while in specifics it is new, the narrow phylum to which it belongs may not be.

Ignaz Maybaum, in *The Face of God after Auschwitz*, distinguishes between two sorts of catastrophe, *Gezerah* (evil decree) and *Churban* (a concept applied to the destruction of the First and Second Temples). While *Gezerah*, Maybaum observes, can be averted by changes in behaviour—"penitence, prayer and charity annul the evil decree"— the wreckages of *Churban*, says Maybaum, "make an end to an old era and create a new era." They are dramatic and dumbfounding— "The *Churban* is a day of awe, of awe beyond human understanding" (Maybaum 156). Neither virtue nor intelligence can affect the course of such an event.

The scale and the rush of inevitability make the species and climate breakdowns of global warming parts of a *Churban*-type event. It thus exhausts the powers of human language and emotion. Quick, confected hope robs us of the formational challenge of feeling this new state of affairs fully. The new sorrow at the heart of this response is an amalgam of at least shocked astonishment, unparalleled dismay and an inability to find durable ground for ontological optimism. No interior paradigms for bearing it exist, yet it would be folly to turn to invention to make up for this absence. Since the calamity igniting the sorrow bites into the possibility of being, it would empty the words ingenuity would rely upon to form its response. Nothing resembles it. Nevertheless, there have been times of extraordinary catastrophe in the last hundred years, where human beings met with systemic, unmitigable loss. Perhaps here we may find adaptable models for how to proceed interiorly now. I dare not argue for one-to-one correspondence between these previous responses to disaster and the current one, but wish to suggest that each is equally a reply to change that appeared to abolish all that came before.

✳ Conventional hope in relation to climate change takes roughly the following form: if the cultural, economic, legal, technological, artistic aspects of a life we now know and move in were reconfigured, perhaps radically, we would continue to flourish. This implicit claim, made frequently enough at academic conferences and international governmental meetings, shows a failure to recognize we presently do and will undergo elemental, irrevocable, substantial loss. How to be in these circumstances is still to be discovered. What is the interior work we must do in order to truly be in this loss? Where shall we turn for instruction? Who recently has been forced to think outside even the faintest consolation of hope, a difficult, inhospitable place for thought about interiority to be?

The Holocaust and the Great Depression forced novel hermeneutics and schemes for living within the events that crushed persons affected by them and in the wake of these events. Philosophical theologian Emil Fackenheim and personalist philosopher Peter Maurin, in collaboration with socialist journalist Dorothy Day, in the midst of their devastations or the residues of these, outlined practices that attempted to begin a relevant thinking and plausible enduring in the face of a previously unexperienced scale of loss. Their specific disciplines include a refusal of all transcendent political, cultural and religious forms (Fackenheim) and a program of performing works of mercy both to other humans and to the land (Maurin, Day).

✳ There was little Jewish philosophical-theological response to the Shoah in the first two decades after the Second World War, Emil Fackenheim observed, because "a well-justified fear and trembling and a crushing sense of awesome responsibility to four thousand years of Jewish faith…has kept Jewish thought, like Job, in a state of silence" (*God's Presence* 70–71). Beginning in the late 1960s, however, a series of thinkers emerged from this silence: Ignaz Maybaum (*The Face of God after Auschwitz*, 1965); Richard Rubenstein (*After Auschwitz*,

1966); Emil Fackenheim (*God's Presence in History*, 1970); and Eliezer Berkovits (*Faith after the Holocaust*, 1973).

Fackenheim, as befits the author of *The Religious Dimension in Hegel's Thought*, was intrigued by the compatibility between suprahistorical presence and actual historical events: *Geist* operates through causal chains of occurrence; indeed, in a sense, it could be said it takes embodiment from them. Fackenheim's entry into historical-spiritual reflection in *God's Presence in History* is not, however, his reading of the *Phenomenology of Spirit*, but the midrashic tradition, where a less rigid but more dramatic form of historical causality is on display. Judaism, Fackenheim notes, is filled with visionaries—Isaiah, Daniel and, perhaps the greatest, Ezekiel, before whom "[t]he heavens were opened" and "visions of God" were presented to sight (Ezek. 1:1). These extraordinary moments were exceeded, though, by an earlier, more demotic experience. Fackenheim cites Rabbi Eliezer, who asserted in a well-known midrash that "what Ezekiel once saw in heaven was far less than what all Israel once saw on earth" (Fackenheim, *God's Presence* 3). The visionary saw not God but "visions and similes" of divinity, while at the crossing of the Red Sea, even the lowliest maidservant saw with immediate clarity the nature of the surpassing presence before the people. "As soon as they saw Him, they recognized Him, and they all opened their mouths and said, 'This is my God, and I will glorify Him'" (Exod. 15:2). Following the Holocaust, Fackenheim argues, the God of history may not be abandoned, though to remain loyal to such an entity raises towering problems of theodicy. These, however, cannot be allowed to turn Judaism into a mystical, otherworldly religion, Fackenheim insists; any committed, thorough transcendentalism grants a victory, he claims, to those who oppose the religion.

It is irrational, of course, to argue that anything approaching equivalency exists between post-Holocaust theology and burgeoning thought around interiorities appropriate to the age of climate change. But in Fackenheim there is an attempt to preserve authentic interiority in the face of widespread collapse, and his insistence that this innerness be rooted in historical memory and acted out in contemporary events is a stance those who attempt to think within the catastrophe of global warming should hold in mind. The shock caused by the effects of contemporary climate problems easily could catapult us from the flow of time into imaginative other homes. Climate change occurs within the causal structure of history; an appropriate response to it is not to leave history. Various escapist transcendentalisms have appeared in the last ten years as ways, one suspects, of coping with the new sadness and premonitional terror, some technological (the ambition to live on Mars, committed pandemic immersion in the meta-world of social media), others have been political and economic. All globalisms, continentalisms, nostalgic nationalisms and committed residences in alternative realities, "second lives," are refusals of the hard historicism of the problem. Most technological, political and economic transcendentalisms are enervatingly Hegelian in that they place hope in forces that, like Spirit, are in history but not of it, not significantly touched by it, as they exercise ultimate control over it. Such forces include the supposed inexorability and benignity of technological advance and an economic evolutionism that, it is claimed, replies to it and is believed to be operating as an organic power behind the emergence of more massive trading and governing forms. A response to these misunderstandings is a Fackenheimian refusal to engage with such meta-worlds, which are ways of evading the bearing of the full novel sorrow arising from extinctions and extreme weather.

✻ The Catholic Worker movement appeared in the United States as a lay, leftist engagement with the Great Depression. It grew from a collaboration between journalist Dorothy Day and French philosopher Peter Maurin, whom Day found waiting in the kitchen of her New York apartment on December 9, 1932, the day she returned from covering a communist hunger march in Washington, DC, for the journal *Commonweal*. Communist and leftist Catholic friends had suggested to Maurin that he and Day "thought alike" (Forest 101), so he had sought her out. An intensive six-month conversation between them ensued, which resulted in the inaugural publication of the *Catholic Worker* newspaper and eventually the establishment of a worldwide network of houses of hospitality and farming communes. Considering the social and economic conditions in 1930s North America, Maurin and Day proposed a three-part ameliorative plan, involving roundtable discussions in various rented spaces in New York on topics as diverse as "Cultural Interests vs. Business Interests," Scholastic theology, Jewish spirituality and racial justice; houses where it would be possible to practice the corporal works of mercy (feeding the hungry, giving drink to the thirsty, clothing the naked, sheltering the homeless, visiting the sick, ransoming the captive and burying the dead—cf. Matt. 25:31–40); and the establishment of "agronomic universities" (Forest 144), where the unemployed would produce their own food in a work environment where the division between labourers and scholars would be erased. In implementing this program, Maurin and Day insisted that all efforts be kept on a small scale and that the communities maintain a persistent tolerance for failure. Day, a Benedictine oblate, also encouraged *stabilitas*, a staying where you are, identifying the gifts and needs of a particular neighbourhood and shaping your work to them.

The works of mercy, from the point of view of outsiders to the Worker life, seem to be pure acts of charity, relatively ineffectual in a larger socio-economic sense. Day, however, understood them to be part of an ascetic practice, a counsel of perfection, perhaps especially suited to times of collapse, when events can make some available and tender. They aided in the construction of the soul, were activities that aligned the self in the world and minutely altered the world in such a way that a larger community of care began to appear. "An act of love," Day observed, "a voluntary taking on oneself of some pain of the world, increases the courage, love and hope of all" (5). The same was true for farm labour at Maurin's agronomic universities: it formed one for a truer placement in being through physical labour shared with various others. Conversation, the third element of Maurin's *paideia*, helped locate the self in relation to others while introducing the larger mind of shared talk into the life of the community.

Catastrophe for Catholic Workers provoked a going to ground in the form of small community life, the growing of food, basic conviviality, trust in the seemingly extra-personal genius of conversation, the discipline of stability and a daily structure of the performance of the works of mercy. Each of Maurin's three activities, together with Day's later adumbrations, comprise a form of rational psychagogic behaviour in response to great public experiences of *catanyxis—we have taken a wrong turn*. These, along with chastened absorption in what Fackenheim calls "root experiences," events that obliterate or massively change forms of reality, are how some may naturally act within widespread breakdown.

None of these interior (and glancingly political) *behaviours is efficient*. Dispositions drawn from Fackenheim's theology and Maurin-Day ascetical activism cannot forestall or physically mitigate what is coming upon us in global warming. The disciplines drawn from Fackenheim and Maurin-Day will *do* nothing in a widespread sense; they, together with practices associated with autochthonicity and the epistemology implicit in the lyric poem, however, can help build an inner stance, a dispositional architecture, which may give us a better chance of holding truthfully and justly the weight of the new ontological and political sadness stealing upon us and may help us engage the sensibility within this novel interior state in colloquy.

✽ Keith Thor Carlson, in *The Power of Place and the Problem of Time: Aboriginal Identity and Historical Consciousness in the Cauldron of Colonialism*, his account of the roots of Stó:lō consciousness in the Fraser River drainage, argues that a tribal memory of a succession of disasters extending far back in time, including stories of the Great Flood and accounts of famines, meant that the smallpox epidemic in the late eighteenth century, originating from trade networks connected with European traders, while "unprecedented in terms of its biological nature, was not incomprehensible in terms of its demographic effects, nor was the social response without Indigenous precedent" (92). Other accounts of collapse in the deep Stó:lō past, according to Carlson, include an unidentified apocalypse that reduced tribal membership to a single individual in the community of Yarrow. This woman, by good fortune, was joined by another sole survivor from another band, who had travelled from the other side of Sumas Lake. The two married and began the rebuilding of the community. A way of proceeding within the cataclysm of smallpox was latent in the historical continuum of the River People.

This chapter revisits some of the recent history of catastrophe in the West and elsewhere, and responses to it, to vivify lost memory of patterns of possible endurance, imagination, virtue and formational practice that previously have functioned within disaster and which may serve as emergent contemplative models in the era of global warming. But this is only the beginning of our exploration.

INTERIORITY AND CLIMATE CHANGE

WE NEED BROAD, UNACCUSTOMED INSIGHT into how to live in a new inconceivable sadness, beyond even what the examples of Fackenheim's hard historical theological realism and Catholic Worker activism can offer us. There is a deeper rooting that is possible, and we must seek it out. I imagine an extensive maieutics, coming from the ancient and the medieval world, a rich part of the Western sapiential tradition, that can help us take up the spirito-political work ahead. Versions of many of the interior exercises I have in mind are, in fact, at work in both Day's and Fackenheim's undertakings, her Benedictine stability, his interest in midrash. The greater range of formational endeavours, ascetical renovations, reliefs, involve the practice of certain forms of conversation, imagination, of thinking and acting and the practice of the discipline of contemplation's justice. Some of these spiritual exercises from the past are considered below. Collapse can be the porch of great contemplative cultures, and we seem to be on the verge of one now.

Conversation

Philosophical conversation, also identifiable as spiritual direction, friendship of the soul, contemplative teaching, contemplative listening, employs many arts. What Plato calls "dialectic" is not abstract philosophical theatre, but a series of intensive readings of particular selves, combined with a simultaneous reach for deeper truths about reality. While "drawing the soul higher," as Simone Weil observes in "God in Plato" (70), dialectic is also a method of conversation that seeks "to attain to each thing itself that *is*" (Plato, *Republic* 532a–b). A theurgy of clarifying thought exists in such exchange, building to an idiosyncratic freedom where the self springs free from convention or despair; in each interlocutor the identification and unleashing of their longing is intended to end in grasping "by intellection itself that which is good itself," this undertaking aided "by discussion" (532b). Such vivifying conversation is rooted in the particularities of certain souls alive in certain times. Now we enter the time, perhaps the early stages of the terminal time, of climate change. How are we to be? What will help us be this way? What will this new ascesis created by these times make us?

I have come to suspect that two apparently different sorts of people will become crucial as the effects of global warming swarm upon us—gardeners and spiritual directors. The value of a gardener's knowledge is self-evident, while that of the spiritual director, the contemplative conversationalist, is less clear. Both soil work and discernment traditionally have been pillars of the contemplative life. The losses caused by climate change, acidification of oceans, collapse of fish stocks, species extinction, massive fires, droughts and flooding, leave us dazed, mourning and driveless. We will need to be reconstituted interiorly and fed back into the world as we live into the coming changes. Quick, depthless, palliative hope here will be worse than worthless; part of contemporary interior healing will be removal of this type of hope from our lives, even as we continue to pursue those goals—dramatic reduction of fossil fuel use, elimination of coal as an energy source, preservation of forests, among other practices—on which this broad, implausible hope of escaping climate catastrophe rests.

Features of a contemplative practice appropriate to global warming can come from the array of Socratic interior therapies or pedagogical devices, the formational tale, the location and cultivation of personal philosophical eros, certain sorts of conversation, among other methods, and from spiritual exercises found elsewhere like *lectio divina* and Ignatius of Loyola's "representation of place," where one's inner savouring of a milieu in which the numinous has been housed—Israel in exile and in the wilderness, say—animates the person within by engaging the contemplative imagination and appetite. By this method, we delectate the scene we hold in resting thought, letting it become, in some sense, us.

And our interior, visual and affective imaginations must be expanded in ever more novel ways these days by representations of previously unnoted places of epiphanic possibility, especially those more rooted in our particular locations. We will find such prompts in, for example, Philip Kevin Paul's poetic vision of pre-contact W̱SÁNEĆ in his books *Taking the Names Down from the Hill* and *Little Hunger*; in a dreamed anti-colonial, anti-anthropocentric, pan-ontological "larger conversation"; in fresh readings of Dorothy Day's and Peter Maurin's acts of mercy and their agronomic universities; in the ontological correction, the alteration of the reading of the terrain, that comes when one's local place "opens its eyes" as one attends to it with care and uses the true, pre-colonial names for things and places as courteous and vivifying words. All these exercises provide routes of affection, meaning, loyalty, identity and endurance. Similarly, we must read the patterns desire is shaping in us. Helping us discern these forms, and assisting us in clarifying their visualizations so they can be masticated within, are parts of the director's work, central to philosophical conversation.

Imagination

Another theurgic device that may become useful to us in the coming years is Zoharic: a practice of expansive innovative exegeses of texts and interior states that aim to build to ecstatic, inflamed construals of one's situation in the world and further possibility and are capable of such animation even if the world declares out of court these creative, extravagant models. This hermeneutical behaviour differs from Ignatius's representation of place in its extreme interpretive reach, the intent of which is to embolden its exegetes and their auditors to interior and imaginative vocational audacity. A theurgical performance like this is at work, for instance, in Louis Riel's *Massinahican*, which he composed while in exile at a small Catholic mission in Montana in the early 1880s and continued through his imprisonment following his resistance, a theosophy that came to him when most of what he lived for appeared to be lost. We have only fragments of this stunning reading of the nature of the world and humans in it, a numinous materialism, an erotic soteriological monadology—but with daemonically charged gestures and words, small bites are often sufficient. Fragments of this vision can be found in his *Collected Writings* and in my masque *The House of Charlemagne*. One sees yet another rescuing theurgy in the capacious, newness-welcoming, saturated listening to the unlike of the nineteenth-century nun Sara Riel in her small convent in Ile-a-la Crosse in what was then Canada's far North West Territories.

A similarity in the transformative hermeneutical spectacles of the three, Louis Riel, Sara Riel and the Zoharic community, is that their ecstatic readings of cosmology, interior intimations and the Torah respectively are altruistic and revolutionarily political: the exegetical reach in each was heightened, and its purpose was equally vast. It was the unification of heaven and the world, resulting in complete healing of all things, the reuniting of male elements of divinity and being with female ones, a conciliation emphasized in both Riel and the Zohar, a meditational practice having, by inches, a truing ontological effect. Here we have a massive but not exotic dream: each of the three pursued a metaphysical realism and believed some version of this union of behaviour and transcendent excellence was precisely what the times required.

Cognitive spectacles like this are tonally non-mainstream, non-university, non-Enlightenment-shaped speech and do not appear in conventional descriptions of thought's range, but their anomaly status sheds light only on the poverty of the intellectual culture sheltered in post-secondary education and elsewhere now.

The first moment in the transformative exegeses filling the Zohar is disarrangement, as it is in the moment of liberation in the tale of Plato's cave. While the instant is painful, it is nevertheless instructive, relocating. Melila Hellner-Eshed, in *A River Flows from Eden: The Language of Mystical Experience in the Zohar*, describes the stages of this heightened interpretation as the companions of Rabbi Shim'on enter into it in their after-midnight gatherings.

> The next stage [after the explication of *peshat*, the simple meaning of the verse and the citation of previous conventional glossings of the text], which serves as a preparation for the mystical ascent, involves a heightened contemplative effort superseding the known, usual and obvious interpretation. This contemplation weakens the control of common sense in the interpretive act. It heightens instead a rarer sense, permitting a higher level of association in seeking the verse's gateways to the mystical homily. The exegetical enterprise is predicated on an acute and powerful set of intentions, the focus of which is not merely another understanding of the text possessing a discursive meaning alone (as the Talmudic saying puts it, "the Torah speaks in the language of human beings"). Rather it seeks to expose the divine world hidden within the holy text, as well as a connection with this world. (331)

There is a shock, "the shock of beauty," in seeing what Weil calls "the material source of an energy directly useable for spiritual progress" (81), the nakedness of matter, which is for her the "real presence" of the numinal. Undoubtedly, this nakedness will be the surprise of pitched individuality, distinctiveness. In the *Republic*, in the Allegory of the Cave, this dumbfounding occurs near the end of the violence the former captive experiences within the boil of light once she is led from the mouth of the cave. The disarrangement comes from the unusualness of the things standing forth before her in searing truth. You see it, arresting essence—grass moving in wind just a little up a slope—in a such a way that it seems to see you. This reciprocal action is a home, a double in-homing; you provide a home for the thing by recognizing its beauty; it provides a home for you in its ampleness and the mercy of its beauty. All this would be a Zoharic result. Such perception amounts to a reconstitution of the self, as well as the mounting of a broader world. Minimally, disciplined attention, an exegesis of alertness, makes space for the nakedness of the grass to appear. One can train another to attend in this way through demonstration and encouragement. Here is where the expanding, retrieved, apokatastatic soul, hastened by a variety of maieutic gestures, turns to the world as home and terminus of care.

Currently these extravagant, rescuing construals exist chiefly in poetry, in Peter O'Leary's *Earth Is Best*, for example, or in Roberto Harrison's *Yaviza* and its exploration of hidden Indigenous solidarities that he calls the Tecumseh Republic. These covert gatherings in his politico-metaphysics are "ghostly" ontologies. Harrison, who writes he has "always thought of myself as an Indigenous person" (*Tropical Lung* 21), spent his early years in Panama, emigrating to the US with his parents when he was seven. The single Tecumseh Republic "inhabits the living and the real. It grows from a Spanish foundation of illuminated words and adds union to dispersal" (*Yaviza* 161). While "mostly inarticulate and sometimes painfully vacant," the Republic, spanning the occluded noosphere of North America, creates a citizenship where each gives "to the hills for the seas to become me," thus "linking the sky and the earth together" (161). These solidarities include "every Indian language and all languages—sensed and beyond—

natural and artificial

of the forest
of the plains
of the desert
of the swamps
of the Sea
of the river (157)

"I am a Tec," remarks Harrison—

"archaic technologies are mine to dream
into morning" (157–58)

Harrison is active in the Indigenous community in Milwaukee, participating in ceremonies (*Tropical Lung* 20). There he dreams the way of the Tec. "I as a Tec abolish the glass/and see faces in everything" (*Yaviza* 169). "I as a Tec repairing archaic wandering" (168).

✱ The extreme outcomes of climate change provoke panic and feverish activism, along with obstinate denial; these responses block elaborate, deeply engaged readings of land, texts, eros's twistings and therefore are to be resisted, treated as heightened, acedia-driven distractions to be held only in peripheral view. Zoharic and Rielean rapt, daring readings and re-readings feed and encourage the inner life upon which appropriate, just behaviour rests. We should remain faithful to the hermeneutical exercises that build durable, just subjectivities, which alone will uphold a perduring interior and political practice under the new conditions.

The land can be read more vivifyingly if it is touched. In the Benedictine tradition, physical work, often involved in the production of food for the monastic community and its neighbours, *labora*, is the supportive sibling of *ora*, the fostering of an interior calm, stability and the woken apprehension of being. Each power flows into and enhances the other, creating a reverence for things and materiality in general, along with *aura cordis*, an ear of the heart that is quelled, discerning, playful, ample.

Spiritual exercises for the time of climate change, too, will flow from the labour and the political economy issuing from it that has appeared in response to chthonic loss and a sense of impending doom, the jobs and practices we've turned to in helpless response to inarticulable grief and precarity—community gardens, rain harvesting, barter commerce, the construction of a life around the village of a neighbourhood, walking to where one shops, banks, works, has one's dental and medical needs met. These endeavours will call out their sustaining virtues and send us in pursuit of them, capacities for friendship, cooperation, self-reliance, openness to beautiful coherence and peace-making, a disposition of slow, attentive dwelling. These behaviours themselves should be read with Zoharic exegetical vigour.

This work, these endeavours, while practical, and indeed political in a subsidiarian, true way, are also clearly multi-levelled interior disciplines. And they are anagogic, causing, by interiority's small steps, change in local communities and reconciliation with the land. They build quiet, at-homeness, and they draw us back from the enervating encounter with the impossible task of "saving the world," a titanism of vocation likely to trigger a raging despair.

Contemplation's Justice
To talk adequately about and occupy the new sadness associated with climate alteration, one should first come to realize the full value of what is being lost. If it is so, as W̱SÁNEĆ wisdom and cosmology state, that all things—animals, fish, trees, stones—once were people, or at least conative, and even now remain with souls (SHELI), animated in the rhizomic making of relations, capable themselves of mourning, then

we lose even more than we now imagine, a rich, resonant, presently unconscious emotional ecology, knowable for us only as persistent, inchoate pain that will appear in this vibrant assembly's absence. But we gain other mourners, salmon, oaks, deer, bears, ducks, hummingbirds, cliff faces; these join us in our grief in ways we cannot say.

On this sentience, which we cannot picture but must believe exists, on this sentience that we cannot name but must believe is alive in our emotions and imaginations, below the waterline of language, rests the possibility of reaching poetry, orphic philosophies, ecstatic dance, our duendic musics and mysticisms, the wells of colour, rhythm and feeding depth in our cultures. Without this intimation, what enlivens would surely disappear. How perfect of them: the unadmitted souls give us what we need to not be crazy, to be expressively human, as we destroy them. Their self-effacing mercy is a model and a call.

Could a soul freshen in a thing, a being, wake from a kind of slumber, when its correct name is spoken to it, the name it itself urges— TENEXEN (duck), XPÁ I Ł Ć (cedar)? Yes, there is a small stirring that can be hardly made out. Extending friendship here is a political act and a transformative act in our present ontological context.

I argue, therefore, as a matter of spiritual direction and formation, and as a form of political action, for a substantial expansion of language through a use of largely forgotten words found within the Western contemplative philosophies to name at-the-moment lost regions within North American subjectivity and through the learning of Indigenous languages of specific places, languages made by such places. A non-triumphal expansion of self—and a thinning of self through dispersal—will accompany this broadening and deepening of language. It will invite more of the warmth and ingenuity and sustenance of the world, as well as unvisited parts of ourselves, into our consciousness. This expansion will stretch our love, intensifying its particularity, as it broadens our grief. This dynamism feels like justice—this new intimacy is the correct basis of behaviour in relations with other beings—and it is truth; I sense it will clarify us, even as it saddens us more. This freshened sadness is itself justice, captured in the W̱SÁNEĆ notion of ŁELTOS, the unbudgeable chagrin experienced upon receiving what one knows to be one's due fate.

The intent of such contemplative pedagogies in the classroom and elsewhere is not necessarily to draw people to wisdom, though the practices will have this effect with some, so that, more attentive to their interior well-being, they develop a meditational discipline based perhaps on a slow, permeable reading of certain wisdom books (Plato's middle dialogues, the *Tao Te Ching*) or they find themselves attracted to such virtues as justice and compassion and attempt to introduce them into their behaviour. The chief goal of these pedagogies, though, is to make people more alive to the texts they read, ontological and linguistic, and the shapes core, vocational eros makes in them, more courteous toward these readings, more patient and savouring, less eristic in discussion, better listeners, taking in what is said by *ore cordis* (the mouth of the heart). This awakening of the interior senses, the ear of the heart, the mouth of the heart, the eye of the heart (*oculus cordis*), will make a flotational device for us in these times.

Sophistry may be pandemic, as Plato claimed (*Republic* 492a–493a), incapable of defeat, so that learning or soul-formation apart from the disastrous enthusiasms and antipathies of mainstream culture may be possible only for those rare few under the protection of the god. The "great, strong beast" (493b) is active everywhere, spouting and enforcing the common view even in the classroom. Yet philosophical moments, moments of grace, gaps interrupting crushing reductionism, do arrive for a surprising number of people, and they can be encouraged by certain teaching and conversational behaviours.

One goal of contemplative pedagogies, in all maieutic traditions, is the formation of a larger, dispersed, coherent self. The interiorities urged by global warming will pursue their own versions of this project. The whole self is autochthonous—its essence elongated, de-centred, solidified by place—and aligned with the times, uttered by a just appraisal of the predominant longing of the historical moment. Thirdly, this broad ecology of self befriends—that is, at its extreme minimum, does not homicidally attack (Genesis 19:1–29)—the angel, which is, minimally, the agent intellect, the apart, gestalt dexterous, yet startlingly intimate mind often resident in mind. More on the nature of this intelligence comes from a reading of Bonaventure in his *Itinerarium Mentis in Deum*.

Thinking

The angel is the intelligence that visits when an enterprise upon which a communal life depends seems at the point of collapse, as is the case in Genesis 18, as well as at other times, like illness or the pitched attention of contemplative knowing. In Genesis 18, Abraham sits at the entrance of his tent in the heat of the day as three men approach. He has been informed he will be the unlikely ancestor of a multitude of nations. He is childless and very old; his biological line, it seems clear, will end with him. His name is changed by divine fiat, and he is twisted from the norm. Estranged from himself, puzzled, Abraham alone, it seems, recognizes these strangers as "lords" and shows them proper decorum: he rushes toward them and bows deeply and invites them to rest under the oak of Mamre and eat. The extraordinary beings reply: do as you say; that is, follow your sense of what we mean. At Abraham's insistence, they stop and "they ate" (Gen. 18:8) the cream, milk and calf Sarah had prepared. The angel appears when one's way seems doomed; the being announces a counter-position of hyperbolic, peculiar plenitude and escape—"Sarah shall have a son" (Gen. 18:10) from whom will come multitudes—often presenting the auditor with a new name to match the new life, a wild wind of fresh gestalts. Openness, overcoming incredulity, followed by insistence, then feeding the source of insight and resolve precede the angel taking into its mouth what we have grown, what we have made from ourselves. Here is an instance of mind visiting from without. Without connoisseurship, an initial response to an approach of such a savouring, discerning, capacious intelligence is to ignore or drive it off; under more extreme circumstances, the response might be to violate or kill the visitor, as we see in the chapter in Genesis following the one we have considered where the destruction of Sodom and the days leading up to it are described.

Spiritual direction is a cognitive discipline forming a capacity to apprehend beauty or beauty's likely location if beauty is, at the moment, furtive. This interior formation is a development of a taste for beauty in its infinite forms, as Ibn 'Arabi would claim, a discerning taste that distinguishes greater from lesser. This training delivers the Abrahamic eye and the grace it urges, leading one to stand at the appearance of this intelligence, offering it hospitality. Direction further is a nurturing of courage to manage the audacity of the vision of beauty.

Bonaventure, a second generation Franciscan, following Francis and his model of "yearning for ecstatic peace in every moment of contemplation" (*Mind's Road* Prologue, 1), was convinced that with "apprehension, if it be of the appropriate things, there follows *delight*" (II.5). What is "the appropriate thing" in a person, a call, a place, object, task? Bonaventure would say it was that person's or object's "species." He means by this word a quality that is layered, fresh, cohering (Latin *species*, "beautiful form," "beauty") and also its essence or intelligibility, its image in the comprehending mind (Latin *species*, "that which a man sees mentally," "model," "ideal," from *spicere*, "to see," "behold"). A species is a thing's proportion, order and individuality; and it is, in an ontological sense, for Bonaventure, the presence of Christ, Logos. Species, as proportion, is "form, power and operation" (II.5); on the palette of the mind it appears as *specsiiotas*, beauty, because "beauty is nothing other than numerical equality, or a certain relation of parts with agreeable colour. Or else proportion may be considered as potency or power, and thus it is called 'suavity' [Latin *suavis*, "sweet, pleasant, agreeable"] for active power does not exceed immoderately the powers of the recipient, since the senses are pained by extremes and delight in the mean" (II.5). Thus, for Bonaventure, individuality is divinity, for proportion is an aspect of particular things. Deep, contemplative knowledge, knowledge of *haecceitas*, is inevitably ecstatic— "[i]n this way the species, delighting us as beautiful, pleasant and wholesome, implies that in that first species [that is, Logos] is the primal beauty, pleasure, wholesomeness in which is the highest proportionality and equality to the generator [that is, the Father]" (II.8). Resident in discrete things noted with care is the all and its glittering intelligibility, pinpointed, clarified, in idiosyncrasy.

What is the cognitive power alive to this elemental and residually primal beauty? In Bonaventure's view, it is the agent intellect, which is, as well, the Logos, the crisp grasp of idiosyncratic completeness. This knowing, presented as an angelic visit in Genesis 18's allegory of mind, on a sensuous level, is Francis's "ecstatic peace," that state Bonaventure mentions in the first paragraph of *Itinerarium Mentis in Deum*, a condition he himself "breathlessly" seeks (Prologue, 2), an appropriate telos for human innerness. One knows the beauty of the world, its aching, infinite, motile individualities moment by moment, that wealth, by the mind that is also simultaneously them.

Acting

In Bonaventure's illuminationism, the mind within your mind, completing mind, is not yours but still is peculiarly intimate. It is imagination as seemingly infinite, apprehending, creative without caprice, creative as intimating, as it assembles the world, thus assembling the self. With this imagination, knowing is religious, healing, perfecting; these are the characteristics of the existential inflation fostered by the government of this mind. Such mind is also quotidian, unexceptional, non-epiphanous, though you build to it, make yourself ready. By this other mind—the oddness, peculiar gestalt generation, the stretch—you enter your actual self; without this assist, this elsewhere, it aided by personal practice, the mind "lying totally in the [unimpregnated, unlit] sensible world… cannot return to itself as the image of God," or Being entire (Bonaventure, *Mind's Road* IV.1). This estrangement from self is itself usual, the condition of the self taken over by a cartoon, the status quo unless this shrunk state of affairs is interrupted or jostled.

The epistemological therapy required to repair the halted, boxed-in mind has little to do with cognition, methodology or technique, but is dispositional, the creation of a "door" in identity (IV.2). Nor is proper thinking cognition as it is usually understood, but a *peregrinatio* from one's regular, atomic, ambitioned, realist state to the limitless expansion of imaginative grasp, which is the warm breast of divinity and full self. This return is also simultaneously ontological and political repair: one occupies the ideal city in a justly comprehended world. One becomes that city (IV.3). You have restored to yourself "spiritual hearing and vision," "spiritual olfaction," "taste and touch" (IV.3), a complete contemplative sensorium, as well as wonder, previously unsuspected beauty and a link at the level of essence with the world. The logos vital in those so reconstituted flavours and forms all their action (IV.4).

This externality of mind as noesis and act Bonaventure also surprisingly attributes to divine understanding and behaviour. Both the agent intellect and divine knowing are routed through angelic availability and ekstasis. He quotes Bernard of Clairvaux writing to his monk Pope Eugenius III on the dynamisms of providence—"God in the Seraphim loves as Charity, in the Cherubim He knows as Truth, in the Thrones He is seated as Equity, in the Dominions He dominates as Majesty, in the Principalities He rules as the First Principles, in the Powers He watches over us as Salvation, in the Virtues He operates as Virtue, in the Archangels he reveals as Light..." (qtd. in *Mind's Road* IV.4). Thus divinity manifests itself to the contemplative soul as mercy; all relations come through the junction of the floating mind in rich leisure, *pingue otium*. This borrowing of powers by the eternal, however, is alliance or, perhaps, subsidiarity and suggests nothing on the matter of limitation of essence; it is merciful, explicative theatre meant to existentiate ineffability within the range of human interior capacity and to begin the conversation between the ground of being and the contemplative.

But in examining "the essential traits" of divinity as Being, Bonaventure can report nothing to us of what is lost to us in core ineffability: he has, though, laid out an angelology that amounts to a full account of providence, its perpetual, multifarious creative powers. He has described Being's relations with humans (and non-humans and non-sentient life) in examining the charges each order of angel enacts, the range of insights that expand possibility and heal on multiple levels. And if the angel, mind before mind, the messenger, is, in large part, the agent intellect, Bonaventure has described a certain sort of cognitional structure. He has simultaneously produced a complete ontology since his account of these relations is nested within a study of divinity as Being, "the root and name of the vision of the essential traits" (VI.1) of God.

The Trinity, he goes on to declare, an inaugural conversation at the heart of being and beauty, divinity, ontological creativity as spontaneous Goodness, "is said to be self-diffusive" (VI.2). This hyper-benign diffusion is "natural and voluntary, free and necessary, lacking nothing and perfect" (VI.2). Consideration of this dynamism transmutes in the contemplative into ecstatic peace and, I would suppose, resilient courage.

From the pole of infinite imagination or logos, this state, this knowledge mediated by the angel, that comprehension, is one way of accomplishing the feat of ineffability fitting itself into the containments of language, image and narrative to produce the mercy of intelligibility. We move toward this sense, however, not by industry, nature or inquiry but by "inner joy" (VII.5). This interior wayfaring is available even now in the early stages of climate catastrophe. Indeed now it is required as a ground of brave and novel behaviour.

CONTEMPLATIVE PRACTICES, CONTEMPLATIVE PEDAGOGIES

Based on a three-part lecture series offered to faculty and staff at Middlebury College, Middlebury, VT, Fall 2018

Teaching

Post-secondary classrooms, we suppose, are places where information and skills, at various levels of sophistication and complexity, are communicated, this exchange amounting to the transmission of expertise. These same classrooms are, as well, environments where contemplative exercises can occur; in fact, they are one of a few remaining locations where this sort of thought, contemplative insight, might unfold. Taking a contemplative approach in teaching enhances the learning of certain skills and the absorption of information; in an atmosphere of permeable attentiveness, historical, literary and scientific knowledge shines with a liberating newness, opening up new ranges of interior possibility and from this, new areas of knowledge. How one holds truths can determine access to further truths and the depth of these truths.

My teaching over the last thirty years has centred on philosophy and on the composition of poetry and the creation of poetic theory. In poetry workshops, when a student is suddenly and deeply struck by, say, the profundity in Chilean poet Pablo Neruda's whimsy, his own voice and range alter and stretch. The suffusing insight concerning Neruda's style rises in the emergent poet and carries him along into unusually daring experiments in metaphor and narrative, and these experiences make for him a particular writer's life. Such a penetrating realization I think of as a contemplative clarity, an animating light appearing and sweeping through an interior sensibility. Conditions that permit such a grasp can be nurtured.

Let me offer another example of transformative learning. A philosophy student, in a state of readerly availability, is startled by a sense that the conversational dynamics in Plato's dialogues are part of an ever-shifting therapeutic endeavour and not, as they may appear to be, clumsy attempts to develop philosophical systems, and, with this certainty, she begins to occupy a different philosophical universe, with new readings of philosophers like Wittgenstein, Descartes, Plotinus, Eriugena, Spinoza, Weil, Rose and Zwicky lying before her. Philosophy becomes for her a form of living.

If the classroom is a contemplative location, the instructor dimly or acutely aware of this becomes a graceful or rough dialectician—from a maieutic point of view, it can often seem not to matter that much which—and her or his strategies may borrow, in part, from the psychagogy of Socrates and others in the Platonic tradition. These psychopompic devices include what Plato refers to as match-making—here: read these books; I know someone you simply must meet: a student is passed on to other, likely more significant interlocutors, both living and dead; possibly a singular, fecund intellectual partnership, an intellectual marriage of sorts, at some point, with one of them will form. These transformative practices include as well provocations to reconsider exhausted former positions largely by making space for the internal arguments of one's partners in exchange to play themselves out and perhaps transmogrify through epiphany, the consolation of adding to an emergent coherence or through *catanyxic* shame.

Spiritual exercises, found within the conversations traced in a variety of Platonisms—Christian, Islamic, Judaic—and in the Neoplatonisms of Proclus, Iamblichus and Damascius, also involve inducements to stretching personal and political imagination through the telling of vast formational tales—visionary recitals, Avicenna calls them—so that one develops the capacity to make out leapingly novel, yet apt, vocational and ontological patterns. These contemplative pedagogies are means of intuiting one's interlocutor's, one's student's, possible future since the chief dialectical assist is the reading, the drawing forth of, the cooperation with, the shaping of, personal eros, this drive read in the contemplative tradition as elements of a not-yet-cohering Ibn 'Arabian Name of God, or, in a Socratic context, the raw basis of authentic philosophical desire. In a poetry workshop, this often wordless longing marks the earliest appearance of a particular style or voice that forms essence in a writer. In the work of surfacing this longing, the tension between conversation, often informed by *lectio divina*, and the construction of comprehensive explanatory systems is resolved by coming down solidly on the side of conversation. Evagrius and other philosophers of interiority have insisted that contemplative grasp precedes the construction of system and makes its completion wait, perhaps forever. System can amount to the silencing of the psychagogic.

A confession before I continue: sometimes when I look at this proposition—that contemplative or meditational practices can enhance teaching—it occurs to me that I am not really talking about anything new or different, that contemplative pedagogy is simply effective teaching that goes on in multiple classrooms daily throughout postsecondary education. At other times, I suspect I am proposing an approach that is more than a little transgressive and in fact subversive, calling for audaciously new instructional behaviours and dispositions. In the use of contemplative pedagogical elements at Middlebury College, in particular in the Sophomore Seminar I team-taught with educational philosopher Jonathan Miller-Lane in the fall of 2018, it was clear that this method was both novel and helpful to students at a number of levels.

Since 2013, Middlebury has pursued a "mindfulness project" at all levels of institutional life, from instruction to governance and support services. The initial intent of this initiative, I understand, had been to address rising levels of student anxiety, but there was also the hope that a concentration on interiority could provide an alternative pedagogical tone in teaching. My purpose at Middlebury, as I saw it, was to investigate, over the course of a series of three lectures for faculty and staff, a possible philosophical deepening of the college's work in the building of contemplative milieus. A contemplative instruction involves the use of new methods, but also the seeing of perennial classroom dynamics in a new light. The new methods, many coming from ancient sources, that primarily interest me are contemplative practices related to reading (*lectio*), knowing, conversation and the practice of justice. I later became interested in considering interiorities appropriate in the time of global warming. Some Middlebury professors and support staff were convinced students now face a future of perpetual activisms around climate matters, various authoritarianisms and the imperative to decolonize one's thought and behaviour and believed these activisms would be unsustainable over the long run without a developed interior life. Each of the following sections could serve as the basis of separate but complementary meditations contributing to the nurturing of such a life.

Let's begin considering classroom maieutics with a thought experiment. I invite you to picture the following scene. Dwell on it in imagination; take your time chewing it over in the inner theatre we all possess. The situation I sketch is not in the least unusual. A student comes to your office after two weeks of classes. She has asked a few good questions in class. You sense something bubbling in her, which neither you nor she could precisely, easily name, but to you it looks like a mixture of heightened, energy-rich curiosity amounting to emergent identity: the student is trying to work out who she is and what she will do; how will she speak, how write; what is her particular way of thinking, of investigating; what is possible for her? She is also looking for allies—what books should she read now? Who could her interlocutors, living and dead, be? What is her "voice," her distinctive way of shaping arguments, speech, point of view, ends, apprehending pattern? She is sitting before you because she has a sense that you just might have something to say to her that could be worth hearing. There is a mild buzz in the air as the two of you speak and listen.

I understand this scene, not at all uncommon in teaching, to be a contemplative situation; it is, in fact, one of the crucial moments in the doing of a certain sort of philosophy. It is an occasion of possible philosophical midwifery. It resembles instances in Plato's dialogues where Socrates engages various interlocutors or where he himself is engaged—sometimes with near comic roughness, sometimes with a Spartan tenderness—by his own teachers, Parmenides, Zeno, Diotima. In moments like this two-week-in consultation with the curious, burgeoningly enthusiastic young writer, scholar, scientist, musician, actor, visual artist, dancer, teaching can feel like a privileged encounter, where what contemplation knows is supremely relevant. Teaching can be a psychagogic event, a profoundly interior event, in these moments; it frees, befriends and aids inner lives on both sides of the conversation. It can be an art of the soul, an experiment toward wisdom.

Incidentally, picturing this encounter, as we have just done, is yet another of these contemplative arts, one called by Ignatius of Loyola in his *Spiritual Exercises* "representation of place," or in Plato is a formational tale with maieutic powers at work. An Avicenna or Suhrawardi might call our modest thought experiment a miniature "visionary recital," in which not only is truth in the speaker, as well as subsequent readers and auditors, laid bare, the truth of a person's inner life, the truth of their longing, but also the means and energy of pursuing further these truths, the courage to ride the momentum of this longing, is released. So the tale feeds twice; it is doubly daemonic, in the actual event and in a report of the encounter. The initial exchange and later mastications are both contemplative undertakings.

Many would find this strange, heretical talk. What could contemplative exercises possibly have to do with university education? Surely this is a mixing of categories, one having to do with self-help instruction, faith, solipsism, withdrawal from the world, an Eliadian error in scholarship, and the other with rationality, meticulous research and the exercise of an adult will. Others, less skeptical about the existence of contemplative elements in teaching, might wonder how to bring a contemplative spirit to instruction. What would a classroom look like if it rested on, grew out of, contemplative values? What would be the significance for students—personal, social and political—of such an education?

I explore here various contemplative exercises, which I understand to be philosophical exercises in the Platonic, Neoplatonic and Stoic sense of philosophy, philosophy as a way of life, to use the phrase of the historian of the philosophy of antiquity, Pierre Hadot. And I show these exercises arising naturally in a normal course context. I draw these practices from diverse places—firstly, as I say, from Plato and the

Neoplatonists, but also from Christian monastic theologians, Evagrius, Maximus, pseudo-Dionysius the Areopagite, among others, from the Romantic tradition in both Europe and North America, from Taoism and early Chinese poetry, Sufism and what I have learned of Indigenous protocols from conversations I have had with nēhiyaw and W̱SÁNEĆ writers, Elders, language-keepers and ceremonialists over the last three decades. I also benefit from insights arising in the "slow philosophy" movement, especially the work of Michelle Boulous Walker in her *Slow Philosophy: Reading Against the Institution*, and in the writings of such twentieth-century philosophers as Jan Zwicky, Simone Weil and Gillian Rose.

This list of influences is simply my personal and readerly background. All of us come from, and are shaped by, different allegiances and traditions, thus the need for conversation to get the largest conceivable picture of contemplative possibilities. Readers of these ruminations may have been experimenting with versions of these Platonic and Romantic meditational exercises and other contemplative strategies in their teaching or conversation for years, and I would have much to learn from them were I to sit across the table from them, sharing a coffee. I shall propose certain exercises along the way for the reader to try out both as a practitioner of certain disciplines and an instructor.

Reading

The goal of one's reading is typically thought to be a full comprehension of a piece of writing: your reading is meant to help you break a text into all its parts, understand each part individually, and grasp the structure by which these elements are connected. In a standard university setting, students are well advised to read with these objectives in mind since they will be tested on the strength of their systematic comprehension of material, as well as their acuity in critiquing the work before them. Such an understanding is thought to be the basis of professional expertise. But there is another form of reading that I would like to explore here; it has a few names: slow reading, reading essayistically (associated with Theodor Adorno's "The Essay As Form"), "reading for life" (in Martha Nussbaum's *Love's Knowledge*), "reading against the institution" (Michelle Boulos Walker) and *lectio divina*. All of these similar ways of reading are "oddities," to use Adorno's term, in the academy and indeed in the contemporary world, and thus they must be coaxed and nourished in order to flourish.

In *lectio* practice, one reads not for architectonic understanding, nor critique, but in the hope of encountering the piercing phrase, and this hope manifests itself as lively availability. One reads to be read, to be constructed, to be made comprehensible by the text. A degree of emptiness is helpful, deprival. *Lectio divina*—"divine reading"—is the enjoyment of the masticable word, the word savoured in *ore cordis*, the mouth of the heart, and inevitably it is formative, even if minutely, orienting. *Lectio* is the meeting with the word that looks directly at you, a line in a poem, a phrase in a memoir or piece of inspired scholarship; it is the approach of the term or remark that seems to read you, the word that, in a sense, says you, or makes room for you to imitate what it means. *Lectio*, interiorly speaking, is a sensuous reading, where the reader is pliant, biddable, alacritous. Its power resembles the energy of Blake's bit of golden string, any bit of string engoldened by choice, that leads eventually to Heaven's Gate in Jerusalem's wall.

The arresting word may be a figure in literature or history. Tom Joad is one of the many tattered heroes in John Steinbeck's *The Grapes of Wrath*; he is not a saint—note his prison record—but he turns out to be the crucial force in keeping his family alive as they migrate from their blown-out farm in 1930s Oklahoma to hard migrant camps in California. I met him at a challenging point in my life, when, making an early, assisted exit from high school and heading for orchard country in British Columbia from Saskatchewan at the edge of winter, I read Steinbeck's book on the westbound Greyhound. Joad's energy, courage and ingenuity were exactly what I needed to step into this new life. I imagine most have comparable stories to tell of formative figures met on the page. *Lectio* is a form of reading that teaches through delectation. It teaches by the transfixing phrase or compelling figure (Dorothea in Eliot's *Middlemarch*, Hardy's Tess of the d'Ubervilles, Marilynne Robinson's Lila or her John Ames) attaching itself to some existential mote in the reading self and magnifying it or elongating it, so that that feature begins to thrive.

Lectio divina, unlike analytic reading, which is often filled with strain, the labour of sifting and synthesizing, is a reading resting on a fully awake leisure, *otium sanctum*, holy leisure. Hugh of St. Victor says in such a state the mind "delights to range along open ground"

(Robertson xvii). Anselm of Canterbury describes such reading as not hurried, not embarked on in turmoil, but in quiet and thoughtfulness, with no task before it, even that of finishing the book as the reader reads "only as much as...[the reader] finds useful in stirring up their spirit" (Robertson xvii). It doesn't reach anxiously after fact. *Lectio* practice may involve audible reading, uttering the words under one's breath so that one "tastes" them on the tongue as well as in the heart's mouth. The words with infiltrating power, different words for each of us, are luminous and become interiorly intimate.

To read well in the *lectio* tradition, you will need a degree of permeability, together with discernment, a sort of interior connoisseurship, letting you know quickly what to fix on and what to pass by, what has the generative feel of eros and what has not. Indeed, since contemplative knowing is chiefly receptive understanding, a key contemplative trait is such discriminating openness or interior availability. This quality is premised on a type of emptiness, which is a contraction, or an edging to one side, of a version of the self, the driver-seat-occupying self. Where, in a classroom setting, a place where the self is usually trained to lodge itself solidly and assertively within its analytic powers, is the self given practice in the art of stepping aside? What most builds contemplative permeability? One place where this disposition of availability can be quickened is in the practice of the contemplative slow reading we have just described, a reading that waits on a text for its daemonic elements to appear. Another exercise to increase porosity is to be found in a sort of conversation which is non-eristic, non-oppositional, where none dominate, where listening well is celebrated as much a speaking well, where fixed opinion is denied a permanent seat at the table, conversation the conclusion of which is an unanticipatible communal achievement, which is experienced with mild relief, but not a sense of victory—"When work is done/Everyone says/We just acted naturally," as the *Tao Te Ching* (17) says. And one may come to interior openness through being broken somehow—illness, unfortunate occurrence, a difficult move—leading to a receptivity that stands next to sorrow, kenosis as reachability's porch.

The Middlebury Sophomore Seminar students, after Miller-Lane and I explained to them the sort of reading we wished them chiefly to engage in during the term, took to the practice of *lectio* with enthusiasm. An intimacy grew over the weeks between the students and the journals we asked them to keep in which they recorded words and phrases that caught them. While we also asked them to write analytic essays on some of their readings, which we graded, we never perused the student's journals. After learning that his raw *lectio* notes would form an entirely private document, one particular student swung away from us, hugging his journal to his chest. Here, in these pages, the self is in dispersion, the ear expanding.

Exercise
Begin a *lectio* practice and *lectio* notebook.

Possible texts to begin shaping a *lectio divina* practice: The Psalms, the *Tao Te Ching*, the poems of Rilke or Gary Snyder. Read in small chunks and savour.

Knowing

A fourth exercise by which an inner, sensorial attentiveness may be woken, in addition to waiting on a text, dialectical conversation and kenosis, involves the valuing of what Freud called primary process as thought. Freud examines what he calls "two principles of mental functioning," primary and secondary process, at numerous points in his career—"Project for a Scientific Psychology" (1895); *The Interpretation of Dreams* (1900); "Formulation of the Two Principles of Mental Functioning" (1911); "The Unconscious" (1915). Primary process is free and mobile, intuitive, not especially logical. It is the cognitive operation at work in dreams and jokes; it is much of the thought-world of very young, prelinguistic children. It is similar, as philosopher Jan Zwicky points out in her *Auden As Philosopher: How Poets Think* (14ff), to Samuel Taylor Coleridge's and W.H. Auden's notion of primary imagination, except that Coleridge's primary imagination is not entirely free: it cannot fail to be awestruck when presented with what it takes to be the sacred or numinous.

Coleridge, in *Biographia Literaria*, Chapter XIII, describes primary imagination as a mimesis of a force embedded somehow in the cosmos: "The imagination I consider either as primary or secondary. The Primary Imagination I hold to be the living power and the prime agent of all human perception, and as a repetition in the finite mind of the eternal act of creation in the infinite I AM." This force resembles the Tao, say, or Logos: the primary imagination as the creative power of the cosmos as it appears in the mind. Coleridge's primary imagination is an ultimately extra-personal, yet intimate, source of insight, the self's alterity, ingenious beyond the self's capacities but complementary to at least artistic knowing, the "prime agent" of perception. Auden's version of this power, given in his essay "Making, Knowing and Judging," is similar but less mystical. Freudian secondary process is grammatically and logically correct, what is often thought of as the adult mind.

British psychoanalyst Charles Rycroft, Zwicky further notes, urges us to add artistic creativity to Freud's list of primary process contexts. This proposal gets no opposition from me, but I would go even further and suggest that we add contemplation as yet another version of this mode of "mental functioning." Like creative activity, contemplative practice, too, can be inspired, associative, intuitive, unafraid of what seems to be nonsense, not metaphor-phobic, full of play, leaping, mobile, all eyes. Be like a bit of down caught in a breeze, John Cassian instructs the contemplative, an attentive bit of down, mind you, alacritously floating.

In the contemplative philosophies, contemplative epistemologies, of Avicenna (tenth century) and Bonaventure (thirteenth century), the cognitive function of the agent intellect undergoes considerable expansion and receives a remarkable degree of autonomy. You may recall from first-year philosophy that Aristotle posits two powers in intelligence, the passive or potential intellect that receives, retains and organizes the intelligibilities of things and connects these with other things, getting these images from the active or agent intellect. In Avicenna, this capacity to discern and capture images, the asymptotically approached essences of things, becomes an external yet intimate power, "compositive imagination" (Karnes 42), active in dreams and prophecy. You truly *receive* insight, bolt out of the blue. How you receive this and what you do with it is another matter and the work of the passive intellect. For Bonaventure, the agent intellect is the logos, the source and the nature of all intelligibility, as well as its apprehension, so you would treat what it sends seriously, valuing it. Writers talk about these visitations all the time—Lorna Crozier (her "better self" quickening a poem), Osip Mandelstam (a whisper over a distance), Xi Chuan (the historical times themselves writing its own literature through him). We need to dilate, make more liberal our receptivity, the receptivity of students, toward this non-intentioned, crisp, often beautiful intelligence.

It is important from time to time to lift the lock-down secondary process has on the academy, thereby making space for creative play and contemplative dwelling; under such conditions the previously unsuspected set of relations and the unapprehended self often drifts into view. This cognitive state of contemplative dwelling is achieved in part by building openness to the genius primary process is capable of, building a state John Keats called "negative capability," a soft-mouthed receptivity for insights arising as visitations from texts and conversations, no clamouring after fact, no reaching for apodictic proof, no pursuit of renown or argumentative victory. One further device for building such availability, used by some writers, is called "free fall," an exercise on occasion recommended as a treatment for writer's block. Often a writer is silenced because she is over-mastered by secondary process issues—my last book sold in the very low thousands and my advance was a high five-figure sum: what am I going to do? How will magazine x look on poem y? Such confabulations are usually the meat of middle-of-the-night intramural symposia, as one mulls over a writing that has become too intentioned, too will-shaped, frozen in self-consciousness.

Exercises

I propose two exercises to stir and expand an interior receptive state, one the reader can do right now and another that can be picked up in the coming days and weeks. Both are meant to build the muscle of interior alertness. First, the long-term discipline: I encourage you to build on the *lectio* notebook begun at the end of the previous section. Entertain the possibility of an ontological *lectio* and an ontological school of the inner person. Take the impasse you are worrying for a long walk and see if the sought explanatory shapes appear through the agency of terrain. Imagine the genius of others, and the world itself, has words for you. Often their incomprehensibility, linked with perduring allure, identifies these urgings as especially valuable. Let these glimpses hang, if necessary, in their mystery, not fully clear but savourable, a *lectio* florilegium. We will see the usefulness of such a collection of *lectio* observations in the next section when we consider how one might share the fruits of one's contemplative life in the classroom. The second exercise is free fall and we can turn to this now. Take a blank piece of paper and let fall on it what occurs to you, what spontaneously appears at your pen tip, realizing there is no editor here, no judge, no other set of eyes. Let the self that dwells under or away from the day-to-day self speak.

What you write may be instigated by what you have just read or it may ignore this completely and instead float up from last night's dream or a problem that's been nagging or a scholarly endeavour, momentarily snagged, that you have underway. There are no limits. Value discriminatingly what comes; do not doubt that this is a useful undertaking, an instance of a form of thinking welcome in your professional life. In these fragments may be the beginning of an essay, a fresh turn in a long-term, protean project, the start of a new poem, musical composition, a dance, or even the opening of a new trail in one's life. Feel free to revise afterwards; let secondary process provide its invaluable, ingenious assistance.

Free fall, and other spontaneous writing exercises resembling it, may be seen, as well, as the production of a text urged by unprocessed, unedited longing, which itself may become a document available to the perusal of *lectio*. We here are trying to establish relations with the intimate, yet estranged, other mind that in certain Sufi epistemologies is a version of the Angel, whose other name is the active imagination.

To Share the Fruits of One's Contemplation
Conversation that is the reach and perfection of philosophy begins with the uncovering of idiosyncratic eros. Sometimes the student will come, as in our earlier thought experiment, with this eros raw, exposed. At other times, the exposition of a craning eros of another will cause the interlocutor's prelinguistic, erotic teleology to spring free.

One of the entertainments in reading Plato is watching the figure of Socrates mature and change as he spends time with various others in conversation. Early in Plato's *Parmenides* (130a), the reader runs across an embarrassing moment, possibly even a disastrous one, where the neophyte philosopher appears to stumble. The very young Socrates seizes control of the exchange, an extremely ambitious, demanding metaphysical exchange, giving a long, unbridled, uninvited, awkward display of his current thought on the structure of the cosmos in the presence of two august philosophical elders, who happen to be experts on this topic. Proclus, in his later commentary on the dialogue, however, commends Socrates on this audacious move, says he does well by exerting "himself to stir up the divine spark in him in preparation for participation in higher beings" (*Commentary on the Parmenides* II.V.781). Socrates does indeed strut his stuff. But he takes a risk; he might be utterly off; the vigour of his intervention could seem presumptuous. His bold move works, though: after he has finished—and one can imagine a collective holding of the breath with those gathered at this point—Parmenides and Zeno look at one another and smile, and then begin their instruction. Proclus calls Socrates's verbal explosion a form of "theurgy," god work, a transformative philosophical exercise, this stirring up of sparks, a candid airing of one's impassioned insights and hunches. Socrates's outburst aided the argument and brought a fuller, more mature instruction toward him. His behaviour could become a model for us, on both the side of the student and that of the teacher. Most importantly his candor and daring have opened a space in him for learning. After Socrates's speech, his role, as Proclus emphasizes, his post-intervention discipline, is to keep silent, letting his energetic interruption be picked apart and stretched and shifted into larger theories or explanatory shapes. A full and passionate expression of one's views concerning the matter at hand, followed by silent attention, here is presented as a philosophical, an interior, exercise that unintentionally creates an inner availability and a cauldron of possibilities for others; it is not a teaching but a psychagogic act, creating conditions in the speaker by which the speaker's soul, in the phrase of the Alexandrian Neoplatonist Isidore, grows eyes (Damascius, par. 38B).

Contemplata aliis tradere—to share the fruits of one's contemplation—is a phrase, a vocational description, found in Thomas Aquinas's *Summa Theologica*. Presumably, we are interested in introducing contemplative moments in our classrooms because we ourselves have a meditational practice that brings a degree of good into our lives. But how to import the fruits of our practice into our teaching? I think we can say that a personal interior discipline is an important element in letting a contemplative spirit into our instructional work; minimally, the contemplative exercises we use in our teaching will appear less gimmicky or contrived when impelled by a phronetic interiority on our part. Let me offer a few tentative suggestions on how this sharing of contemplative benefits might play out in instruction in order to open up a broader exchange.

First, though, we must grapple with the notion of *disciplina arcani*, an ascesis found in many traditions, the discipline of secrecy. All recognize the dangers of bringing private interior experience into the public sphere. We may say more than we will later think we should have—that's the obvious one. But the more important danger concerns the essential furtiveness indigenous to contemplative insight. Some truths gather force if they are reported, even reported broadly, unrestrictedly, while others evaporate, deflate, become caricatures or become entangled in enervating interpretation. These understandings, often the most intimate and plunging, appear to work best if they are kept veiled. After some experimentation in exposing them, you may prefer not to report them at all. These truths are just for you to consider in private. Candor also can create the unintended outcome of discipleship.

Contemplative pedagogical practice is, then, not primarily a confessional practice, because certain understandings can't thrive in an atmosphere of general perusal, or at least can't keep their original, vital shape and thus their original daemonic charge. And so we lose them by reporting them. Or our interlocutors are thinly fed. But there are situations where the opposite is true. Let me offer three examples of where psychagogic good comes from confession.

Socrates's disclosure of the Pythia's revelation that none was wiser than he (*Apology* 21a), and his vigorous attempts to refute this revelation, make up the substance of several dialogues. Throughout these encounters, he speaks about his extraordinary interior affairs. I think, too, of the twelfth-century nun Gertrude of Helfta's reports of visions in Book Two of her *The Herald of God's Loving-Kindness* and of Ibn 'Arabi's account of the epiphanies—"visitations" of the Angel—experienced as he performed the Meccan rituals recorded in his book *The Meccan Openings* (*Futuhat al-makkiyya* II, 325 in Corbin, *Creative Imagination* 140, 324). All of these disclosures of interior truths offer encouragement to others to believe in the expansiveness of their own inner lives and thus gain citizenship in the world one's own meditational practice causes to form. How to distinguish between times when one should talk and when one should keep silent? Sometimes the anchorites in apophthegmata would offer a "word" to a visitor who had come a great distance, and sometimes they would not. No rule seemed to be in operation, but the readiness and need of the questioner, amounting to a capacity to hear, seemed to be important factors.

However, while confession can't be ruled out of contemplative pedagogy, it nevertheless should be treated with discerning care. We turn to focus now on some less risky, more oblique ways of "sharing the fruits of one's contemplation." These methods too come from the lives of the three contemplatives I just mentioned—they are dialectic or conversation, the telling of formational tales and the providing of a luminous, compelling model, an awakening phenomenology.

The conversations Socrates pursues as a result of the oracle's intervention are all with figures of Athenian society reputed to be wise, statesmen, well-known intellectuals and those renowned for their piety like the naïve and famously self-righteous Euthyphro. Socrates seems to be genuinely mystified by the divination passed on to him, since it is not at all clear to him that he is wise in the slightest. Yet he knows as well that the oracle could not have uttered a falsehood. His post-divination conversations reveal where the celebrated wise have gone astray, misidentifying certain qualities in themselves or where they have been formatively misconstrued by others in certain flattering but unhelpful ways. Their Socratic encounters are occasions of upending, the experience of the sort of shame John Chrysostom calls *catanyxis*, the shattering insight that can shift a life. His interlocutors talk themselves to a point where a gap of doubt or unknowing appears in them, and in this flash of emptiness genuine philosophical practice may begin. But often it does not.

Socrates, it becomes clear, at least has a degree of personal discernment—thus he has a fair idea of himself and can tell in an instant when wisdom is genuine. He also has intellectual humility: in another place, he says that what he chiefly knows is philosophical midwifery, how to bring to term the burgeoning foundational eros in those with whom he speaks. He also has skills as a philosophic match-maker; he can see who should be linked with whom in order to bring interior possibilities to fruition. These powers constitute his understanding of erotics, the foundation of the contemplative philosophic life. You should read Heraclitus. Here, read this long poem "Leaves of Grass," the 1855 version. Do you happen to know the work of Robert Duncan? Dionne Brand? Bejan Matur? Thomas Merton? I really do think you and so and so should meet. Socrates can discern the next step, as well as the misidentification of self in his conversational partner that must be overcome. He has a keen eye for the eros or self-defining, future-creating preoccupation that is not fully known or acknowledged but that nevertheless marks and drives, in a subterranean way, those with whom he speaks. He watches for this power and cares for it once it appears, often by whittling away at it or, to shift images, rearranging the furniture in its rooms. For one who teaches creative writing, as I have done for the last many years, this defining eros almost always manifests itself as a distinctive voice, an idiosyncratic emotional tone and a signature way of using language, but also often as a compelling writerly telos, a sense of mission or responsibility. These Socratic therapeutic powers around the ministration of eros are the traits of a good spiritual director, a SELW̱EN in a W̱SÁNEĆ cultural context. These qualities—an understanding of foundational or core eros, a discerning eye concerning it, a solicitude toward it, a sense of the next step, a sense of the necessary community this eros seeks—make up Socratic wisdom. They are also the qualities of good teaching. Socratic dialectic is not without judgement; it aims at a "therapy of desire" as Martha Nussbaum in *The Therapy of Desire: Theory and Practice in Hellenistic Ethics* describes its work, whereby passion learns an attachment to those things—courage, justice, phronesis, beauty—capable of forming the girder system of a meaningful life.

 Ibn 'Arabi, a thirteenth-century Spanish Sufi, lived in a broader world than most of us do. In addition to our world of the senses, the real, tangible world, there was for him, as there is for most of us, the world of hidden truth, mathematical and balanced relations expressed in planets' orbits and the operation of the brain as it listens to music and so much else, the intelligible rebar of being, what nature loves to hide or veil; and, in addition to this, for him, there was the *mundus imaginalis*, the domain "where the spiritual takes body and the body becomes spiritual," a world accessed by the "active Imagination… the place of theophanic visions"(4), where "visionary events appear in their true reality" (4), as the scholar Henry Corbin writes in his masterful *Creative Imagination in the Sufism of Ibn 'Arabi*. Anyone who has read William Blake, Rumi, Hsieh Ling-yun, St. John of the Cross, Teresa of Avila, G.C. Waldrep or Peter O'Leary has travelled in this world. And Ibn 'Arabi lived in, lived out of, this expansive ontology and wrote out of this multi-ply reality, where the Real was in fact divine, and the form of his life, as he lived in this milieu, became an important teaching and source of liberation for others. In his great book *The Ringstones of Wisdom* (*Fusus al-hikam*) and elsewhere, he lays out a picture of a vast world, an ample formational tale, into which the interiority of an auditor or reader can stretch and grow, the very world he beholds growing within him. Listening, reading, one begins to occupy this world; one's self grows to become commensurate with the expanse of this world. Such a deep philosophical exercise of imagination—imagination inflamed by crisis—may rescue us as selves and nations as we move into periods of cultural exhaustion and breakdown.

Let's look lastly at a third phenomenological possibility, example, as a way of making available to others what one learns in contemplation. The largest part of Gertrude the Great's *Herald* contains reports of others on her presence in the monastery and their inference, based on these observations, on the nature of her interior journey. The contemplative inner life shows its face in behaviour; enacted, this life is read, savoured, by others. In the case of the contemplative in the classroom, this inner life shows itself in a decorum that is a courtesy to those with whom one engages in conversation, a courtesy of attention toward texts and a courtesy, a solicitude, toward the emergent vocation, the forming selves, emergent Names, to use an Ibn 'Arabi formulation, emergent elemental eroses, in one's interlocutors.

One final word on conversation as a philosophical therapeutic: if one begins one's interactions with interlocutors with a report of one's experience within *lectio divina*, the subsequent exchanges stand a good chance of being psychagogic rather than eristic. Talk under these conditions is enhancing, light-birthing, rather than competitive.

Exercises

Begin or deepen or renew a meditational practice familiar to you (this may be a designated *lectio* practice). Begin conversations on texts and events with a *lectio* report.

Give some thought to how you are going to use what you now find in your *lectio* journal in your teaching. There is the possibility, of course, of using in classroom remarks the phrase that has pierced you, or, more subtly, there is the possibility of speaking or acting out of the spirit the new mind that the luminous phrase has placed in you. Or you may choose a confessional route: I read this passage and it puzzled me, stopped me in my tracks. What do you make of it? Was anyone else caught by this? What has been luminous to you may be so for others. Occasionally I will ask students to engage in a free fall exercise on a reading in order to surface penetrating understandings that haven't yet fully come into language.

Allow me a short note on the idea of pursuing a focused, long-term contemplative discipline in a teaching context, the life this may amount to. *Conversatio* (a frequent sojourn in a place, with an echo of *conversio*, to turn around) *morum* (behaviour, custom), an attachment to, or devotion to, alteration, reversal and progress in a contemplative way of life, is a phrase that appears in *The Rule of Benedict* (Benedict of Nursia, Prologue, par. 3,9). Indeed, *conversatio morum* is a notion central to the foundational document of Western Christian monasticism. It represents a fidelity to, and movement in, an assumed interior discipline of perpetual renewal. This commitment takes place within a community and it's buoyed by this community, as the discipline fosters the community. There is a loose, evolving community in a classroom, and one may pledge oneself to interiorly grow and deepen in this; such an undertaking to grow, to be on the way, is the essence of *conversatio morum*. There is a kind of quickening poverty in this; a large and compelling "not yet" stands in your life and propels it.

Contemplation and Place, Contemplation and Justice
It's possible to take the view that contemplative practice, both private and applied to the classroom, if not mistaken (we are all committed to analysis and application of tested hypotheses here), is at least an indulgence. There are unprecedented, urgent political and ecological dangers facing us now, climate change being the chief, and we must act. Anything less than full and committed political and ecological engagement is a shirking of responsibility.

To this objection, I reply with a question. What is the root of politics? I am enough of a Platonist to believe politics starts in peoples' elemental desires, in their consequent ways of understanding the world, and in their very sense of self and how the self might or should be encouraged to develop. I'd like to spend some time at this point looking at a particular version of the self, the self as exile, or as exiled, the self as a divided home and the life of desire and the politics that ensue from this way of comprehending the individual.

The sense that the self is split and that there exists a diasporic, extra-personal second self is an ancient one. It appears in the story Plato puts in the mouth of Aristophanes in *Symposium*, where the tale serves as an account of the origin of desire. The first humans, recounts the playwright, were beautiful, gigantic spheres composed of female-male, female-female and male-male parts. Zeus had these wonderful beings divided because of their hybris, and thus unintentionally he created all-absorbing non-religious yearning. The ancient sense of self-division appears, as well, in Teresa of Avila's account of her auditory mystical experiences, her locutions: another mind, a second better self, close but autonomous, addresses her, and in her interior marriage with this other conversing presence, she becomes whole. Hegel's *Phenomenology of Spirit* concerns this untreated sense of self-separation and how time and talk bring momentary relief and finally an awakening to the truth that all persons might be connected at a fundamental, deep identity level, at least fused in their momentum toward a shared, essential goal, the acme of spirit. Repeatedly in the *Phenomenology*, the same drama is played out: spirit, constrained within a limited point of view, encounters another view, another Weltanschauung, and sees it as oppositional, but as it continues to engage with this view, it comes to see the new understanding as augmenting itself, while requiring change in one's initial self-consciousness.

As the Hegelian dynamic continuously unfolds, the alert philosopher, phenomenologist of spirit, stands at what Gillian Rose calls "the broken middle" where the "drama of misrecognition" unfolds (*Mourning* 72). Hegel calls this state "the bacchanal of the spirit" (*Phenomenology of Spirit* 27) and "the speculative Good Friday" (*Faith and Knowledge* 191); delight and sorrow mark the place. The philosopher of the middle stands in this troubled intermediate place listening, asking questions, discerning linkings, sustaining the creativity of the dialectic that is no single person's achievement but arises naturally from persistent, perduring exchange.

The ultimate and true Teresan, Aristophanic and Hegelian selves are larger than we usually suspect the self to be; this larger self is achieved when an apparently alien shard—a peculiarly personal "other" or series of others—is met or retrieved, without undermining its unlikeness. Our usual self in North America is the atomic, self-determining, self-asserting Millian self, the version of self defined and defended in J.S. Mill's *On Liberty*. But this anthropology is not ample enough, it seems to me, not audacious enough, sufficiently imagined, to be true. We North Americans are much psychoanalyzed, yet our levels of anxiety remain high. Indeed, they appear to be rising. This makes me wonder if there might be a deprival, a lostness, at a deeper, chthonic band of the psyche that present psychology has difficulty reaching.

    A feature of modern life has been a pattern of deracination, found in cultures built on both colonialism and mature capitalism. "All that is solid melts into air," under the rule of the bourgeoisie, observed the impressed Karl Marx in his *The Communist Manifesto* (476). Could the full self be in fact a larger ecology of person intertwined with local place? One sees this hunch explored in Wordsworth, Thoreau and John Muir. More recent bio-regionalists like Denise Levertov, Brenda Hillman and Aldo Leopold also, in various ways, inquire into this possibility. W̱SÁNEĆ people, on Vancouver Island and the San Juan Islands, speak of persons belonging to a particular stretch of river or a span of shoreline, rather than the private property other-way-around version. I wonder if part of our malaise might be an ontological loneliness; our severed, lost part may indeed be a place, the place in fact where we are now.

    Some Canadian thinkers have been exploring the matter of autochthonicity, the grace of coming from one's ground, for years—see Mark Dickinson's *Canadian Primal: Poets, Place and the Music of Meaning*. I first ran across the concept in philosopher George Grant's essay "In Defence of North America" (in *Technology and Empire*), the notion growing out of a train journey Grant made to the Rocky Mountains in the early 1960s. He was astonished by the scenery, as most people are when they encounter these peaks for the first time, but he was skeptical about whether he was able to receive the full charge of what stood before him. We may recognize, he wrote, that there are gods here, but we know they can never be ours, because of what we are and what we have done (Grant 17). What we are: we have lost touch, Grant believed, with the contemplative, receptive root of Western culture that would have made us fully, enduringly alive to such beauty. What we did: we met the new land as conquerors and subjugated it, effacing the unlike in what was there that would have shored us up. A sympathetic reader of Grant might suspect that an element of estrangement inevitably stands at the base of our relationship to land.

Lately the difficult, perhaps impossible, task of becoming autochthonous and whole through a vivifying link with land and locale has come to seem even more complex: some of us in settler Canada have entered the historical moment when we see, in our dwelling, we have blood on our hands, realizing that the acquisition of land by diasporic Europeans in the long history of colonization has been marked by both trickery and violence, both within and outside the treaty process, and, further, we see that the land that could be an essential part of our extended self is changing in unhappy ways, or vanishing entirely as a result of climate change, disappearing under rising coastal waters, or being altered beyond recognition as a result of fire summers in the boreal forest. We see the search for embeddedness in terrain—acting respectfully toward it, looking with care, honouring the true old names of places, names that in the Indigenous languages rose from the land— must take on the additional duties of seeking redress for and with First Nations communities, while engaging, at the same time, in a new sort of ontological mourning, what Ashlee Cunsolo calls an "activist mourning" that refuses to let an appropriate sorrow, just sadness, end (Cunsolo and Landman 8–15).

The gulf between settler and Indigenous worlds is significant. Anishinaabe legal scholar John Borrows warns that the task of reconciliation, which should include making a place in the Canadian legal system for Indigenous law on land and land use, will not be a light one. "What we love has been lost," he remarks. "It's been stolen fair and square, with kindness, deceit and a trickster-like efficiency" (Borrows 25). The lost, loved things include not only land but language and the cradle of the ancestral stories as well. "Reconciliation," he insists, "should not always force the Aboriginal interest to 'give way'" (132).

All this, the mourning, cleaving compunction, the hard quest for autochthonicity, seems a tall order, but the pursuit of it is a contemplative-political task essential to take on as a perpetual project. The quest for durable identity, at the very least, urges it. Fulsome, just citizenship urges it. As we move closer to our special place, we see there is also a sorrow inside the land, because the land has history mixed in with it, a history of illegal dispossession, a history of genocide, physical and cultural. The undertaking of *theoria physike*, the contemplation of nature, of place, then, cannot result in the sublimity of Romantic rapture, which, in certain concentrations, is a form of amnesia concerning the lachrymose residue of history embedded in land, but must include remorse, *catanyxis*, and a commitment to engage in an ascesis of truth and engagement with First Nations.

But maybe, just maybe, this difficult task of coming from one's place is not the nearly impossible burden it appears to be but in fact, viewed via another sight line, is a relief and consolation. Iamblichus, the great Syrian, theurgic Neoplatonist of the third century, taught that the human soul, among all other forms in the hierarchy of life, gods, daemons and so on, possessed the highest degree of "otherness" (*heterotes*) (Shaw 101), thus, unlike gods and daemons, it was not strict, inviolate, in its autonomy, but was given to identify itself with a range of what was other than itself. The soul also, Iamblichus believed, was predisposed by its nature, again unlike other presences in being, to undergo the pain of self-division; in fact this elemental, troubled leaning defended the soul's identity, since the soul preserved its identity "by always changing its own essence" (Shaw 104). For human interiority especially, Heraclitus's maxim is apt: by changing, it rests. Being rendered, disarranged, practicing "loyalty to loyalty" in the language of Josiah Royce, always reaching outside ourselves to particular causes and communities, we return; leaving the atomic self, we come to meet our fully expressed selves. The discomfort in this reaching is part of the package of the human vocation in the cosmos, in Iamblichus's view.

Contemplation is a matter of going home, returning to the self's true nature, returning to the cosmos as ordered and beautiful, a state of justice, even if it exists only as a heuristic, restored. Interiority is depicted as this homecoming in Taoism and in Platonism: return to the genius of water in all its forms, return to the unifying principle of being. Contemplative practice, thus, is apokatastatic practice, the practice of returning to an original wholeness: our nature is such that we are inclined to this healing of return, of restoration. I lean toward an ascesis of the recovery and use of a true, non-erasing naming of terrain to facilitate a return to the land that, for various philosophical and political reasons, appears lost to us, an extension of self wrenched away. This ascetical practice, the recovery of original Indigenous names, may cause the land to look at us and, to a certain extent, take us in. There is also the discipline of taking our poverty into the forest and setting it before the trees—T̯EWEN̯ QEN̯ON̯ET TEWE (Have pity on us; look at us and care for us). Then there is the practice of taking on your place as home.

Thoreau's *Walden* is full of prompts for assuming the larger body of place: "I went to the woods because I wished to live deliberately, to front only the essential facts of life, and see if I could learn what it had to teach" (85). "Simplify, simplify," he adds a few sentences later. The entry to the contemplative life can be easy, a single look. "For many years I was a self-appointed inspector of snow storms and rain storms." Leave the desk and put yourself outside and pick up "the discipline of looking always at what is to be seen" (105). Practice this slow reading of landscape in the mood of sacred leisure: "Sometimes, in a summer morning, I sat in my sunny doorway from sunrise to noon, rapt in reverie, amidst the pines and hickories and sumachs, in undisturbed solitude and stillness, while the birds sang around or flitted noiseless through the house, until by the sun falling in my west window, or the noise of some traveller's wagon on the distant highway reminded me of the lapse of time. I grew in these seasons like corn in the night" (105). He gathers the harvest of an ontological *lectio divina*—"Every little pine needle expanded and swelled with sympathy and befriended me" (125). The land looks at you; things open their eyes before you.

Exercise
Explore a contemplative life as the root of a pursuit of justice. Think of the interiority art-making demands as another form of contemplative practice and activism.

HOPING FOR
SOMETHING TO APPEAR

The Poetry of Don Domanski

In Memorium[1]

Don Domanski repeatedly took himself to magical, liminal places in his poetry, along "the edge of a forest/that is capable of everything" ("Edge," *Selected Poems* 7). He went to such places as an intermediary, a metaphysical adventurer, and brought us with him out of our cramped Cartesian enclaves where everything is shabbily too bright and over known. And thus he placed us in touch with energy and mystery, and he was a refugium and healer since he acknowledged, befriended and moved in a multiplicity of worlds.

Domanski's worlds aren't the product of capricious invention, but the recovered essential elements of the *real* one, the "vast earth" familiar to contemplative philosophers like Ibn 'Arabi and the people who left their art on the walls of Chauvet cave over 30,000 years ago. Because he lived as an artist in this varied place, he was under-celebrated in a literary climate favouring a one-ply realism. And because of his recognition of a multi-layered reality, chthonic and metaphysical together, he was a rare poet capable of changing lives. No one else wrote like him. He has left us with such plenitude in his books, and all of it strikes me as not just poetically thrilling but psychagogically solid; if he shows us stairs to ascend or descend, we know we can trust our weight to them.

A great burnisher of the ordinary, Don Domanski made us look at the details of our days with a scintillating freshness. In line after line, the supra-real clusters around unattended ordinary objects and processes. He spies the remarkable from his own middle-of-the-night perspective:

> Outside the snow is
> turning away. Outside the dark leads the snow
> down to the harbour for second life across the
> expanse of ice. ("One for an Apparition," *Earthy Pages* 7)

Reading passages like this, we, too, bump up against a sentience, a *conatus*, at play in things, a force that is funny, purposeful and intimate. We are dislodged from a reductive, object-indifferent seeing and liberated into a broader, more interesting, ontological citizenship. It takes courage to be this different—*atopos*, unusual, out of place, Plato would say—both as a writer and one who looks, and from the very beginning of his career, in *The Cape Breton Book of the Dead* (1975) and *Heaven* (1978), his work was bold, reaching further.

✳ I first met Don in the mid-1990s when he came to Saskatoon to read at two poetry series I then ran, one in the city and the other at St. Peter's College a hundred kilometres to the east. After picking him up at the airport late on a freezing Sunday night, I took him to an otherwise unoccupied Chinese restaurant downtown and we plunged into a conversation that ate the hours. We discovered that we shared class roots, he growing up in rough areas of Sydney, Cape Breton, I in northwest Regina. Somehow we'd both found a way to poetry through, in my case, gangs, recreational crime and horrific fights.

I don't know where in his past he may have found the exquisite gentleness with which he treated the small and theurgic things he noted, his teasing and courteous way with divinity-infused rain, fox tracks, scarf-wearing fleas. His poetic treatment of these presences made them pop out of the flat face of normal ontology, the greyness of bland regard, and become mythic and mighty, central to the parliament of all that is; they hold worlds; they *are* worlds.

✱ Tim Robinson, in *Listening to the Wind: The Connemara Trilogy, Part One*, observes how anglicized Irish place names dry out and degrade Irish places, leaving them open to exploitation "for lack of a comprehensible name to pinpoint their natures" (qtd. in Thubron). Domanski managed to make out unacknowledged powers and to retrieve his degraded things through close observation, lavish imagistic invention and music, and so revivified a more ample world capable of sustaining us, making the self both more effacing and larger. The roll of his language, the lushness of image coming after image, the musicality of his thought sang through, undulating, the attentive reader. His music was at times even more piercing than his jolting imagery.

In my mind, Domanski was after that presence in objects and events, an innerness that, post-Kant, was said to be unreachable, maybe not even there. As he moved from cranny to cranny in the ground, in the air, deep in the grass, in the ghost-world of rivers, to the wolf's "swampy breath" that "presses at the door" ("Wolf-Ladder," *Selected Poems* 84), places unexplored in our present culture, he followed animals and insects, their tracks, advice or charisma without hesitation. He followed rain to "the held pulse of the soil," "all spiders this side of Elysium" ("A Thin Place," *Selected Poems* 277), he followed bears and coyotes. These, in turn, became singular, reliable guides to further rambles into the inscapes of earth, but how to locate, then approach them? With the stretching permission imagination brings to knowing, the daredevilry of the presumptive leap. Halfway through the arc, the arc of metaphor or narrative audacity, of Domanski's leaps in being, you realize that though you have nothing beneath you at the moment, your gamble with these spirits will pay off, indeed that your travel to these otherwise inaccessible corners is the only course open for you. You are brought to "the black river under the earth," face to face with other numinal flashes of "the spirit of other things" ("Osprey and Salmon," *Selected Poems* 161).

Credo ut intelligam, wrote Anselm of Canterbury: I believe so that I may know. Faith makes for large noesis. All true. But so does imagination, the knotting of wildly unlike things—unlike yet still calling out to one another—in image and narrative unfolding, the setting aside, or ignoring, of the propriety of old categories, following some interior hunch concerning the vastness and complexity of creation. Reading the connections Domanski made, one dwells in a feral, more merciful, bigger-girthed landscape because someone has dreamt it, named it, peered into its individualities and that dream has entered the reader.

The small hours were Don's anchorage and his time of resolution, of *quies*, that rest and "negation of wounds," where a silence may appear "passing over us unseen like a dark forest/scarcely a centimeter or so above our heads" ("A Thin Place" 277). He conjured this resolved world repeatedly in his work, and this practice made that world real, inhabitable, a precarious state that one could dwell in by often going there, following that particular spiritual exercise. What breathtakingly ingenious generosity there was in this labour, extending the gift to all of us, the thing or power we hope to see "bending/the meadow grasses ever so slightly" (280), its yearned-for gaze directed at us.

✸ He liked to add quotations to some of his correspondence; one note sent to me on October 30, 2011, ended with a paragraph from twentieth-century neo-Thomist Jacques Maritain.

> Metaphysics enjoys its possession only in the retreats of the eternal regions, while poetry finds its own at every crossroad in the wanderings of the contingent and singular. Metaphysics give chase to essences and definitions, poetry to any flash of existence glittering by the way, and any reflection of an invisible order. (Maritain 235–36)

Don's reach for the unlike was a form of knowing, "a pre-verbal reality, a calling forth from the core within our being" (296), as he puts it in his 2005 Ralph Gustafson lecture "Poetry and the Sacred" at Malaspina College. He was all for the "invisible order" Maritain mentions, and because his allegiance to that occult republic was utterly serious, he must have known the wildness of his thought and creating placed him in some peril, the danger known by the insurrectionist.

With American poet Donald Hall, he saw poetry as a "revolutionary act" in a culture committed to burying the sacred.

If this is true and poetry is numinous sedition, what is the state that emerges if Domanski's epistemological insurrection, poetry's insurrection, is successful? You could say that new place looks like the works of Don Domanski. It is a world in which athletic imagination is valued highly, the gnosis that allows you to "fly over language," as he said in the afterword to his first selected poems, above "the ruins of custom and interpretation, mighty edifices meant to last millennia" (*Earthly Pages*, Afterword). In this new place of imagination, we are near the Kingdom of Heaven, rescued from the asphyxiation of individualism and the tiny room of the present insistent preoccupation, in league with the dead—his dead including Eckhart, Rumi, Hildegarde of Bingen, Lao-tzu—and "a deer's porcelain footprints" and "a vapour of rabbits in the hills" ("Nocturne," *Selected Poems* 283). Here, in this broader community, our heart's need for wonder, a hunger Domanski knew well, is appeased.

✱ Domanski was a generous friend to many writers, generous in reading work and commenting on it, generous about setting us on our feet when we toppled, and generous also in his gifts. Even there, in friendship, he had a love for the far-away and resonant—like the natural cast of rain drops from the Lower Carboniferous from 350 million years ago that he gave me some time in the late 1990s. I, in contrast, was much more fumbling in my gift-giving, carrying back from a trip to western China a ring I'd purchased in a bazaar a short walk from Ta 'er Si, Kumbum Monastery, on the Tibetan Plateau, in old Amdo Province. I had taken the ring to be silver and of deep Buddhist significance only to be told a few days later by a gleeful Domanski that the ring was not silver at all and bore the image of the Ka'ba. He was as gentle with hopeless gift-givers as he was with his small creatures.

I do not let myself think that I will never read again poetry fresh off Domanski's nocturnal desk. I can't bring myself to take this fact in. Last summer, a few months before his death, he sent me a poem[2] from his weem of Halifax's middle of the night.

> tonight the totality of all beings weighs
> less than a feather

less than the movement of eyeshine
radiant in the thicket
less than the weight of tomorrow
balanced on a single blade of grass

I love nights like this   hour's quiet
and driftless   the biosignatures
of invisibilia filling the air   close
as flesh to vein   as vein to earth
close as the cuddle death of a queen
deep in her hive.

***

this night appears in the green world
almost existing   almost not existing
like subatomic particles
like the prehistory of a sigh
like the grace of sentience   spellwork
of consciousness throughout the foliage
it emerges from somewhere between
the intimacy of insects
and the forgetance of men.

Notes

1. The title of this essay is a line from "Sunrise at Sea Level," first published in Domanski's *Heaven*. This chapter, in a slightly different form, was the foreword to Don Domanski's *Selected Poems, 1975–2021*, published by Corbel Stone Press, 2021.
2. "Small Hours" reproduced with permission from the Literary Estate of Don Domanski.

POETRY'S PRACTICE OF PHILOSOPHY

Anne Szumigalski

The Anne Szumigalski Lecture, League of Canadian Poets AGM, Saskatoon, SK, June 8, 2002[1]

How dear to her is the journey of the mind,
flying from dwelling to dwelling,

Her feet scraping the tops of the
forest trees as she floats on by,

Exchanging one language for another,
never quite sure of her bearings,
counting the chimneys on unfamiliar roofs.

—Anne Szumigalski, excerpt from "Theirs Is the Song"[2]

SOME WOULD SAY Anne Szumigalski saw it wrong: the life of the mind is simply not livable in the poem. And some of those saying this would be poets. The poem is the theatre of feeling; the poem is the pool of light where passion stands and sings its termagant songs. They might concede that intuition could sit snug and happy in the poem, imagining this as a sort of feral knowing, somatic, far closer to empathy than dialectic. So not thinking, never thinking: the poem is where you go for haven from the tyranny of abstraction. The poets who say this might be surprised to find that they had what looked like allies among the philosophers, who would declare that poetry is subjectivity to its very bottom. But poets shouldn't be fooled by this appearance of alliance: there will be no free meals for them from this source. Philosophy doesn't love the poem construed as the house of subjectivity: it holds it in the same contempt as it holds metaphysics, while recognizing its utility: here is the isolation ward of hysteria.

While philosophers may have such private views of poetry, they almost never have gone on record with them. I can think of two exceptions to this near total silence. Martin Heidegger, of course, seems to refuse the norm, offering poets priestly status in the temple of being, but the even more famous example is Plato's expulsion of the poets from Kallipolis, the beautiful city, the city of philosophy. I think we misunderstand these two philosophical utterances on poetry, Heidegger's and Plato's, well-known though they may be; and I further think that by unravelling these misunderstandings, especially the one concerning Plato, we'll come to a fresh comprehension of poetry that will show it not antipodal to philosophy, but sitting alone in front of philosophy's hearth, tending the fire.

Poetry and philosophy. I can imagine few prominent, mainstream philosophers who seriously think *Leaves of Grass, Lyrical Ballads*, or even *The Divine Comedy*—to say nothing of *Life Studies*; *Birding, or Desire*; *Songs for Relinquishing the Earth* and *The Beauty of the Weapons*— are philosophical books, places you might look for light into the darkness philosophy explores. In fact, to so believe is to accept marginal status in the world of professional philosophy. It is now far from obvious that poetry, religion and philosophy are linked in any way— their split is virtually an article of faith: one way to think about the last four hundred years of European thought is to see it as a focused effort to detach these three endeavours and to denigrate two of them, poetry and religion. But Homer didn't see things this way, nor did the person or persons who wrote the long poem of the descent of the goddess Inanna to the underworld, nor did the Haida poets Ghandl and Skaay—in the work of all of these, the three undertakings go on simultaneously; indeed, they make a single doing powering the poetry. I think Ghandl and the others were right: poetry, philosophy, religion come from and return to the same place: contemplative attention.

Listen now, for a moment, to Julian of Norwich on prayer. This is from a middle chapter in her *Revelations of Divine Love*. The religious language may be a shock to the ear, but try to hear the form of what she says, try to take in with the seeing ear its erotic shape.

> For prayer is the means by which we rightly understand the fullness of joy that is coming to us: it is true longing and sure trust too. Lack of the happiness which is our natural lot makes us long for it. Real understanding, love, the recollection of our Saviour enable us to trust. Our Lord sees us constantly at work at these two things. It is no more than our duty, and no less than his goodness would assign to us. It is up to us to do our part with diligence, yet when we have done it, it will seem to us that we have done nothing… (127–28)

> But when our Lord in his courtesy and grace shows himself to our soul we have what we desire. Then we care no longer about praying for anything, for our whole strength and aim is set on beholding. This is prayer, high and ineffable, in my eyes. The whole reason why we pray is summed up in the sight and vision of him to whom we pray. Wondering, enjoying, worshipping fearing…all done with such sweetness… (128–29)

So, prayer as desire, as, more specifically, erotic passivity, a doing nothing with implacable resolve, the what-is-not-the-self stippling in. Waiting, and something coming into the waiting that is not moulded by anything, not even anticipation. Here is the doppelganger of the poem that is mere obedience to what presses the tongue, in which one writes best by doing the least with a strenuous, an elegant, commitment.

Let me stray back to Plato. I'd been saying that we don't really understand what he says in the *Republic* about poetry. We read the dialogue usually, as Iris Murdoch did, as sharp-toothed antipathy between Plato's cerebral puritanism and sensuality of all kinds, poetry's in particular. We're wrong, I think, to come at it this way. Another reading of what's going on in the *Republic* around poetry shows that poetry and philosophy are one.

Bear with me; let's walk down the basement steps of the dialogue into the cavern of the book; it's badly lit with a poor fire; you smell mouse shit and wax from guttered candles as you dip your head and enter. It is, of course, night. People are talking; the room floats on a living wave of talk; the talk is sometimes rhythmic, sometimes jerky, like the movements of a large, possibly dangerous, animal. There are, say, ten people in the room: some are in love with one another, some soon could be in love. Socrates, Plato's teacher is here, so is Glaucon, Plato's brother. They're in a room in a seaport not far from Athens—the rotting sexual smell of the sea enters and falls apart around them. The air, if we could see it, would be darkly green, slowly growing. Socrates and Glaucon have just taken in a religious festival, horses ridden hard on the beach by men carrying fire; the room twitches with numinal energy. Dionysus is on patrol here; be careful, keep your eyes open.

Glaucon is the man of the hour; the talk all eventually flows toward him. He's an interesting specimen: he likes disputation and has a taste for political grandeur—if he were alive today he perhaps would have a job in the Bush White House; he's imperial, eristic; he loves war. His name might be David Frum. He's a rather dangerous personality: charming, ruthless, charmed by his own limitations. He got all of this from poetry. Poetry's ruined him: the erotic deformity of his soul comes from an entranced reading of a particular poem, the *Iliad* of Homer—the giddy ferocity of the assault has stained him—and when Socrates hammers away at poetry in Book III and later in the dialogue, he's trying to free him from this book and the fascism it's nudging him into. Suppose he succeeds—he doesn't, but suppose he does—and Glaucon suddenly feels the ax-cleave of shame for his political hybris, what's left for him? Well, aside from mathematics and astronomy, there's that other book, about one of the fighters at Troy and his return home. In the ruin, the kenosis, the romance with the celestial lover, the descent to the dead, the apokatastasis found in the *Odyssey*, Glaucon would read of the shamanic path of Odysseus and the pattern of philosophy, and he might even internalize and begin to enact this pattern. His eros then would unbend.

Just what is philosophy? I'm not that great a fool that I would attempt to provide a full answer to this question. But I can chatter away for a while about what it roughly is in the *Republic*: there, philosophy is a particular unfolding of desire: a sorrow awakens eros in the non-desiring man; a lavish self-emptying goes on; there are tears, a perpetual disposition to tears; there is the travel below the ground and a final homegoing. A turning around of the soul, as Socrates puts it, entirely erotic, entirely resisted, entirely desired, utterly refusable; it is a turning of the soul so that it finds itself loving the things that return it to itself. The work of philosophy here is a close version of the ceremonial work of Odysseus—intensely private, yet altruistic; psychic, political. This is what is advertised as philosophy in the dialogue; no one really takes it up there, but it's on view in the shop window. It is a journey replicated in foundational poems from many places, as I've said—"The Descent of Inanna," Skaay's "The One They Hand Along"—and it's reproduced, in the ancient and medieval world, in contemplative philosophy in dialogue after dialogue, and in tract on mystical theology after tract on mystical theology.

This same sort of erotic unfolding happens in certain sorts of contemporary poems too—not necessarily long poems either, though it doesn't fear length. The work is partly about the long walk of unsatisfiable longing—*epektasis*, the endlessness of desire, as Gregory of Nyssa called it. The poetics of the poem I most admire and which I wish to write are really a philosophical therapeutics, an ascesis.

I've said it comes down to contemplative attention: this is what's at philosophy's heart, at the heart of poetry, the centre of religion. Philosophy, of course, doesn't do this sort of thing any more, and religion, for the most part, is transfixed by cosmology and television. Poetry alone, as I've said, still visits the old fire site. But what is contemplative attention? It is what happens to you when you are knocked to the ground by some astonishment. You go very still at some point in yourself and become entirely eye. What can do this to us, hold the attention in such a way that the Odyssean transformations are required, quickened, that the interior travel—the tears, the descent—begins? Lovers, sex, landscapes, God (I think of Pier Giorgio Di Cicco's new work), great horror (I think of the work of Peter Dale Scott, especially *Coming to Jakarta* and *Minding the Darkness*). And when you look hard at such things (at political horror, at sex, at God)—as you must if you truly see them—your life comes apart, just as it begins to drift together. You are entering the cloud. And not only do you break up, but language begins to do so as well. It becomes besotted like you, woundedly drunk.

Here are two sections from a long poem called "Kill-site." The man you will meet in the first section, Henry Kelsey, was the first European on the plains. He took a two-year walk in the last decade of the seventeenth century from Hudson Bay that brought him as far west as the Touchwood Hills, about 250 kilometres east of Saskatoon.

Kill-site

The animal dreamed of me,
a brown gust separating above its head;
this was below snowhumps on the creek,
ice fog up the towers along
the valley sides, blood on the snow,
estrus marks, the water frozen a yard down.

When Henry Kelsey died or left
Hudson Bay, there's a rumour he continued walking under the ground
in the highest part of his voice, down the west hip of the Porcupine Hills,
a pythagorean thrum in his eyes.
Because all this was a new music, uncooked ratio, a machine of smoke.
And he thought *Let the will sleep here 400 years.*
Let the will sleep here 400 years.
Only some song would turn the lock.
He was looking for the deeper Crees, Naywattame, the Poets, people sleeping
    along the rock ledges behind their eyes, someone to put something
in his mouth.

And because he was under the ground, everything came to him—he
saw a face of wheat, a face
of mineral beam, nipples of stones, a face
of winter in things, face of what is
at the back, the watery, the alto part of the mind,
showing through skin.
(It rose to the skin
    for its hump to be seen, then moved back into the trees.)
Only a song would turn the lock.

He kept walking, wanting and fearing
the freezing of rivers.

Sandhills in a light, likely-daylong rain, looking off
  to the left, grass that's not going anywhere, September—everything walks
toward you; it undresses and comes
        toward you with its small bright hands
and the downwind smell of your father's mind
and his shoulders in the early summer of 1964,
he's working two jobs, post office, moving company; right
now he's not wearing a shirt, a hundred and forty-five pounds,
but still less under the name of his lower-in-the-throat citizenship,
where he's not saying a thing, living in a cave two
thirds up a cliff line, how
did he get there, swallows heaving in front of his face, the hole
trench-shovelled into clay sides lifting over the Milk River, north of
the Sweetgrass Hills, cattle clouding off infinitely
to the east, feathers
and bones hung from string at the mouth of the cave,
pale green feathers smooth out long and speechless from his tailbone.
Things climb out of the elms of their names and themselves
and they come forward, moving their tattoed, Fulani hands.
They smell of your father's voice, his
    one
      black
        suit.

Martin Heidegger said ontology was first philosophy; all thinking grew from an account of Being. Emmanuel Levinas, his student, sensing a threatening Hitlerian inflation in the tone of this project, nearly losing a wife and daughter in the Holocaust, losing relatives, said no, no, ethics was the originary thinking: everything grows from the dumbfounding before the other; a profound courtesy is the residue of this astonishment. I have an idea for a third possibility—erotics, mystical theology, as first philosophy. I suggest this standing before the spectacle of the failure of dogmatic theology, the mammoth racial arrogance it has underwritten, the hybris of a particular culture. What is elemental, quintessentially desirable, is beyond knowing, speech—to presume to know is to begin a colonization of the feral world, to pay the world the great compliment of likening it to yourself. Who can say the large thing which desire wants, wants, wants? But we can watch eros as it cranes for it, see the poses it takes in this reaching: these shapes are adequate ideograms for what cannot be said, and reading them doesn't ossify behaviour, but builds it, plucks it.

I'll come right out and say it, the thing that roils beneath all this attempt at explanation: poems are an amorous pull, a being in love, that plush, mine-shafty chaos, and sometimes this being-in-love feels like a sickness. So poetics is erotics, is mystical theology. Where's the poet in this? He's keen and useless; he's passive, ready for anything; he's ignorant and busy with the heavy labour of opening, which is, in part, a refusing, a divesting.

A word of caution. One given over to eros's doings likely will become what those who followed Socrates called *atopos*; that is, unlike, discreditable, laughable, possibly feared, significantly weird—her eye is caught helplessly by, sequestered by, the out-of-range. But then this liminality draws a certain darkness, slant, commodiousness into the poem that is otherwise not available to writing.

In Heidegger, or at least in the essay "The Origin of the Work of Art" (*Basic Writings*), the poet is the passageway through which the world speaks its truth, the poet is the single priest of Being. If the heroism of high Romanticism, the heroism of the Heideggerian poet, strikes you as implausible, if the narcissism of this figure seems repellant, what can you think happens as the poem appears? The way I see it, the poem is far more than the writing of it, and the best language to talk about this far-more comes from contemplative philosophy or ascetical theology. You fall apart before some arresting thing, some terrible beauty, and you empty. If you stay low, this thing may come toward you like an animal from the forest. This permeability before astonishing otherness, and what this astonishment makes you do, is also philosophy. Such language—the language of rapture, of psychagogy—has no standing, though, in modern philosophy, as I've said throughout this chapter; it has not had for many, many years. It appears there as delusion, the betrayal of the four-century, pan-cultural project of reason that has brought us democracy, the MRI machine and more television than you can shake a stick at. Poetry conceived this way, as contemplative attention and availability—a life given to this—seems a dangerous backsliding in relation to this project. Poetry isn't marginal; it must be made marginal: it speaks heterodoxy—just by the way it goes about its business. Good for it.

I want to return, finally, to Anne Szumigalski. Here is the whole of the third section of "Theirs is the Song."

> How dear to her is the journey of the mind,
> flying from dwelling to dwelling,
>
> Her feet scraping the tops of the
> forest trees as she floats on by,
>
> Exchanging one language for another,
> never quite sure of her bearings,
> counting the chimneys on unfamiliar roofs.
>
> One day she hopes to understand progression
> how it has no end and no beginning,
> how nothing precedes or succeeds,
> how time is a disc that wobbles
> as it spins.
>
> The melody is an old one
> played again and again.
> All night she's aware that it scuttles
> over the pillows like a louse on a holiday.
>
> Waking she hears it emerge from her nose,
> a hum like paperwasps.
>
> "But that's just the tune," she says,
> "tomorrow on my way I'll write the words." (3)

How lucky we were to have Anne Szumigalski's mind, her eye, with us for a while. Her life was a rare visitation, one of the few things we can hold up to someone who might step from the horror Europe has inflicted on the New World, who might step from capitalism's eating of the planet and demand an accounting, one of the few things we might be able to present to a possible survivor and say, here, this was done well.

Notes

1. A version of this chapter was previously published in *Measures of Astonishment*, University of Regina Press, 2016.
2. "Theirs Is the Song" reproduced with permission from the Literary Estate of Anne Szumigalski.

READING WILLIAM CHITTICK
READING IBN ʿARABI

I SIT IN A CREAKING CABIN near the shore of Last Mountain Lake in southern Saskatchewan as winter is about to begin. The morning light is cold and milky. Thousands of sandhill cranes rustle in harvested grain fields around the narrow one-hundred-mile-long body of water, the lake where, incidentally, my brother and I vacationed with our parents when we were children, where I got terrible sun burns that eventually gave me melanoma. Most of the cottages nearby are closed for the season, boats winched from the water. The cranes feast on spilled grain left by combines before they resume their journey to the Gulf of Mexico along the great North American flyway. I am reading William Chittick, who is reading Henry Corbin and others reading Ibn ʿArabi (1165–1240), at a kitchen table, salt, pepper and sugar containers pushed to the edges. I take careful notes and stumble through the strange, expansive cosmology.

The cottage has only a small, underpowered heater to warm it, and I've been asked by the owner to use it sparingly, so the mornings and nights are cold. My mother is close to death in a care home in southern Alberta, in the town where my brother lives; I have purchased a cheap cell phone and only my brother has the number. I travel with my one suit in case I must attend my mother's funeral, which will be held in Regina, with internment in the Soldiers' Plot, as close to the grave of my father as possible. In a few weeks' time, rehearsals will start in that same city for "The House of Charlemagne," a dance in three acts based on the life of Honoré Jaxon, a performance dreamed by the Métis artist Edward Poitras, choreographed by Robin Poitras of New Dance Horizons, for which I have been asked to write a text. The afternoons in the hills are mid-September warm; there are snow flurries some mornings.

✽ Ibn 'Arabi was a wanderer, a person who fed on retreats and visiting students of the way with whom he believed he must speak. In 601/1204, he was in Mosul, in 601 he was also in Jerusalem, in 602 in Konya before returning to Jerusalem. In 608/1211, in the midst of a thousand-mile pilgrimage to Mecca through the Maghreb, the philosopher and mystic from Murcia, in Spain, had a vision near Bougie, in which he "saw himself united in marriage with all the stars in heaven and then with all the letters of the alphabet" (Addas 178–79), saw himself become consubstantial with the cosmos and the language it urged.

✽ Chittick begins his treatment of Ibn 'Arabi within the context of a larger examination of the Sufi forms of knowledge with the observation that "[s]omewhere along the line, the Western intellectual tradition took a wrong turn" (ix). This error—largely an over-emphasis on a certain calculative reason, together with a commitment to a severely modest empiricism—has resulted in the loss of access to what Henry Corbin called "the imaginal world," *mundus imaginalis*, where "objective body to *intentions* of the heart" is given (*Creative Imagination* 224). Louis Riel's *Massinihican*, his almost entirely lost metaphysics, is an instance of such a world, numinous materialism choked off by the colonial endeavour. "Once we lose sight of the imaginal nature of certain realities," Chittick claims, "the true import of a great body of mythic and religious teachings slips from our grasp" (ix).

✴ I am reading in this cold cabin a twentieth-century exegesis of a twelfth-century Islamic Neoplatonist because long ago my saintly mother, who had only her grade eight Depression-era education, woke in me a yearning for spreading understanding by having long, floating conversations with me in the kitchen on Dewdney Avenue, when the rest of the family was at work or had returned early to school. We talked about deeper, personal, elemental things; though I can't recall our actual topics, I remember the sensation of intellectual satiation I took away from these sessions.

Ibn 'Arabi's vision in Bougie, him growing living connective tissue with the heavens and the alphabet, forming a single matrimonial body, puzzled him, and he asked a friend to communicate the details of this apparition to a local interpreter of dreams, while keeping his identity hidden. The interpreter, impressed by the dream report, declared: "This is the bottomless sea. Whoever had this dream will receive a portion of the celestial sciences, of the hidden sciences and of the mysteries of the stars and of the letters which nobody else has obtained in his time" (Addas 179). He continued, "If the person who had this dream is in this town, he can only be the young Andalusian who has just arrived," and the dream interpreter expressed a wish to meet him. Upon hearing this, Ibn 'Arabi began immediate preparations to leave that place— *disciplina arcani*.

✴ The apex of Ibn 'Arabian epistemology is infusion, a range of knowing available to the Sufis. His word for this sort of insight is *futuh* or "opening"— synonyms are unveiling, tasting, witnessing and divine self-disclosure. "The prophets and the friends among the Folk of Allah have no knowledge of God derived from reflection. God has purified them of that. Rather they possess the 'opening of unveiling' through the Real" (Ibn 'Arabi *Futuhat al-makkiyya*, *The Meccan Openings* III 116.23, qtd. in Chittick xii). This form of receptive comprehension, one could suspect, is close to Martin Heidegger's *alētheia*, the revealing "unconcealment of beings" (184, 198), except that in Ibn 'Arabi's case what is disclosed is the Real, that is, divinity in the form of the world and its specificities, each one of the infinite Names, one of the infinite relations between God and being, while Heidegger, at least in "The Origin of the Work of Art" (*Basic Writings*) is not theological in this way. But on closer inspection, Heideggerian "unconcealment" and Ibn 'Arabian "unveiling" have an even greater difference, appearing at the start of each's epistemological practice: in Heidegger, one receives truth by simply being an "artist," a "poet" (whatever these words may have meant to him), no mention of disarrangement of previous forms of knowing that, as he has it, violated the object, no training for the new cognitive operation required; in Ibn 'Arabi, the ascesis of retreat, of discipleship and invocation must be the life of the one who wishes to "taste" (Chittick xii). One approach, Heidegger's, produces a Romantic fantasy that lacks discernment from the moment of initial self-dramatization onward, and in it one is vulnerable, as Heidegger himself was, to a shift to the hard right in politics and its *übermensch* mysticisms. The other track comes to an ecstatic metaphysics, where the kenotic self becomes a participant in the substance of being. A third and final difference lies in the fact that the Folk of Allah enjoy the assist of the Quran, which is the "actual, true, authentic, embodiment of God's Speech" (xv). The Book "provides the God-given and providential means whereby man can come to know things in themselves, without the distortions of egocentrism" (xvi). The Book, thus, itself aids in the beneficent reduction of the possible knower. This self is lean, attentive, phronetic, the Heideggerian puffily "heroic," stagey, easily fooled.

✳ The intimacy with being found in Ibn 'Arabi's Bougie vision echoes the cognitive intimacy suggested by the word "tasting" (*dhawaq*). The world, as it is, appears on the interior tongue. What changes to the world, our view of it, what changes to us, would make such closeness possible?

"Finding," Chittick observes, "renders the Arabic word *wujud*, which, in another context, may be translated as 'existence' or 'being'" (3). Such a grasp, as "finding" suggests, is an apprehension of the physical world, the cosmos or the spectacle of the relations within the physical world as a theological experience. "The famous expression 'Oneness of Being' or 'Unity of Existence' (*wahdat al-wujud*), which is often said to represent Ibn al-'Arabi's doctrinal position, might also be translated as the 'Oneness' or 'Unity of Finding'" (3). Finding or tasting is a knowing that takes place within a state of confusion, Chittick adds, that is "not the bewilderment of being lost and unable to find one's way, but the bewilderment of finding and knowing God and not-knowing Him at the same time" (3). This peculiar lostness is the state of being in possession of true knowledge; "tasting" God is in part being God, at least that fragment which is my Name, fully being the Divine Attribute or relation which happens to be my particular essence, without fully grasping it, my thisness, a Name that had some of its first moments in me in those long lunch-hour talks with my mother in my very early teens.

Cranes, Last Mountain Lake

The party's over around the weedy, lanky
water, jet-skis tipped on dinky beaches,
boats of pirate flag flying cottagers, watermelon
heads, welders
at the steel plant, being winched, unroped
from algae.
Above Regina Beach lair smell—
deepfrying; oleumed, handled twenties,
biker money slipping into cabin cruiser
laminate, the car wash, black
stone cladding firms—
sandhill cranes are in
fluid orbit.

The guide out of the rain
flares a mouse nest of prairie wool
with struck flint in one of the fireplace
chimneys in a cold shed
at Issac Cowie's fur post, historic
site unvisited, cars streaking a hundred yards away to
Regina,
burning ball twisted
above her tattooed wrist as it balloons
exhaling into riverstone dark.
A man at Pelican Point
told her Louis Riel held séances in Métis
cabins along arms
and points of Last Mountain Lake
on his sleepwalk to Batoche

and there breathed out
the army of the plains, buffalo
hunters' deer-hide feet in stirrups, conducting
them onto mares to mill
against the short queen,
which followed him north.
Here's his number.
March, horses slipping in slush,
the lake, trapped mauve dusk, frozen
solid as sky
fibered with returning snowgeese and trumpeter swans.

*

Song metal filings, air
horn flecks,
sandhill crane voice matter, puffs up from alkali
slough fringe
and swathed hard wheat circles spinning
to the shore all week, but no
vision of lordly birds,
until the night I pack
the rented Nissan, when northwest wind,
from bison commons' floors in Wood
Buffalo National Park, Great Slave
Lake, thousand
stickless miles of glacier-
dropped stone
punts the cranes into twisting pillars,
slings them to
eastern Montana then skips them
to the Gulf of Mexico.

*

Cowie, the guide
continued, evaporator of herds
save hides through the Cypress Hills
and down this valley and the next, smallpox
inoculator, trimming wafers of lymph

gland from Pascal Breland's
daughter's neck to brew
warmth in limbs in Fort Qu'Appelle
slept years
on a single plank behind the chimney
stone the lit grass
in her hand lightly heats.
Outside frost-haired poplar walls
last birds levitate from wind strop-
whitened stubble,
wobbling into shaking air.

✻ Early in his commentary, Chittick recognizes Henry Corbin, especially in his *Creative Imagination in the Sufism of Ibn 'Arabi*, along with Toshihiko Izutsu (*Sufism and Taoism: A Comparative Study of Key Philosophical Concepts*) as the two most remarkable twentieth-century hermeneuts of Ibn 'Arabi. Chittick, however, finds shortcomings with both engagements, Izutsu's lying in his preoccupation with abstract metaphysical notions and his exclusive concentration on the hermetic *Fusus al-hikam*; as a result, he misses, Chittick argues, an account of the *mundus imaginalis* situated at the heart of Ibn 'Arabi's ontology, as well as "the practical sides of Islamic spirituality" (xix). Corbin's treatment, while it encompasses the mystical along with the philosophical aspects of Ibn 'Arabian thought, misconstrues, Chittick believes, the concept of *ta'wil*, which means to "take back to the origin" or interpretation. In Chittick's view, it was impossible that spiritual interpretation in Ibn 'Arabi would extend past Quranic exegesis: to stray beyond this would be to edge more deeply into Shi'ism that Ibn 'Arabi was prepared to go. Perhaps. But in the *Fusus al-hikam*, especially in its theology of divine names, a hermeneutical approach moving toward the broadly ontological theophanic is inevitable—

> The Real willed, glorified be He, in virtue of His Beautiful Names, which are innumerable, to see their identities—if you so wish to say: to see His Identity—in a comprehensive being that comprises the whole affair insofar as it is possessed of existence and His Majesty is manifest to Himself through it. (Ibn 'Arabi 3)

Corbin was a scholar not only of Ibn ʿArabi but also of many others in the contemplative philosophical tradition. His larger endeavour was to illuminate the esoteric life chiefly as it is found in Ibn ʿArabi, while also tracing its presence in Suhrawardi, Jacob Boehme, Emmanuel Swedenborg and William Blake, among others, where a belief in, a living in, an additional complementary world was central. "The conviction that to everything that is apparent, literal, external, exoteric (*zahir*) there corresponds something hidden, spiritual, internal, esoteric (*batin*) is the scriptural principle which is at the very foundation of Shiʿism as a religious phenomenon. It is a central postulate of esoterism and of esoteric hermeneutics (*ta'wil*)" (Corbin, *Creative Imagination* 78). The literal, in Corbin's view, holds within it the potency of what is beyond. "This potency calls for the action of persons," he wrote, "who will transform it into act, and such is the spiritual mission of the Imam and his companions. It is an initiatic mission; its function is to initiate into the *ta'wil*, and initiation into the *ta'wil* marks spiritual birth" (79).

✳ My mother's *ta'wil* of elm shade and dusk, with prelude:

The Pavilion, the Veranda Circling, Hanging Kerosene Lamps

In the room with the big coal stove, the entire
20×20 house, after the first three girls,
the fourth flu-dead, Evelyn, months back in her arms, you
appear behind a curtain kicked with gusts, your mother's
mucked, peaking face; outside, wheat
in ticking heat reddens
and, when the wind roughs it,
turns to surf, soft machinery.
Florence lives another ten years,
midwife, angel of the area, her sleigh's runner tracks
slicing others' leaving snow-clogged drives nights
and days twenty miles from town.

noon of the potato flowers,
three o'clock, dust-flocked
of ladybugs

Purple tips of timothy through all of the first week of June!
You are in the world
south of the Pipestone River,
silverweed, gold finch's wave-with-
a-hitch over field thistle edges, you are
below the pelican-place, east of the Hart's stone house, near the knoll
above Gooseberry Lake,
where dancebands appear; and in the hall,
with its circling cream veranda and hanging kerosene globes, Fords
parked like a spilled jar of pins along coulee sides,
Mart Kenny and His Western Gentlemen begin to play;
base and snare come through
mosquito's liquid knots (dusk in coulees) and talk
from picnics around the pavilion,
where people eat
devilled eggs, canned chicken, and
your eye begins to build (*The West, A Nest and You*)
and the two stroke hack of the twentieth century
is suddenly audible through poplars;
but still the smell of horses always,
their sour haziness, inner moony-ness of harness
nickel and on particular nights, furniture stacked in the yard,
fiddlers lean into oil light in houses on the odd quarter section
and begin a waltz for neighbours.

No crop, 1934, and they made you skis,
brother and father, from soaked almost-straight barrel staves,
so on them you slick over calloused drifts
roofing four feet of snow over countable stubble,
you with, you say, no real courage,
mother-gone years, working for the Roys or Wilsons,
the stoves and their kindling, bread browning in the high left chambers,
fourteen, you clip on snow climbing to the third barbed wire strand
across fields home weekends, two men
waiting in unsplit silence in their 160 acres,
half still in poplar.

*In a dream
it was as if God were actually
in the skull
a muscle opening like a fan.*

August again, potato crop hopper-stripped,
family living mostly outdoors in lean-tos because of the heat, 1936,
getting by on crows' eggs and trained-in cod,
your brother picks a guitar
made from wiggled-out screendoor wire
behind a hay rake
left in the yard's highest grass, sun
more heat than light in a boil of dust,
barbed wire scar above his lip skim milk
of starlight.
He strips wheels off the rake and fits them on a 2×6
wooden roadster
and misses the Dieppe Raid by the hair's breadth of a fractured leg.

Ocean Man Assiniboine reservation,
in hip-high grass, grazes in wadded heat ten miles
south in The Lost Horse Hills.
You have heard about this
but now are busy standing on your head in a photograph
of you and your sisters, Agnes' thumb holding your dress
to your one knee, all of you shoeless, feet lignite lumps.
The paint on the building behind you horned off by wind.

*It was as if God
were in the head,
a muscle opening like a fan.*

Spring and
over the plains
from Missouri coulees to the boreal forest wall,
aspen leaves click
and drive people into sleep
with tiny silver and green whips.

Year of shelterbelts caving
and leaning in loose dirt.
By August, stomach down, you can lie on the ground
and see the other side of the fields through stalks.
Amazingly, no starvation.
The two mile long grasshopper flail
ripples through dust repeatedly
as the cloud eats closer with its tusks and venting plumes.

What are you thinking?
you ask the elm shade
inside you, by the creek,
in evening, mosquitoes floating
in snarls, mosquitoes at eye height
in the shade, the unsteady eyes of the shade.
What are you thinking?

✳ The forces shaping my mother's young life, the poverty, hard work and mourning, made up the ascesis that permitted the spiritual hermeneutic by which I am convinced she instinctively lived. Her compassion and discernment show me she lived from such a probing reading of the world. The presence of this particular alertness in her life meant her spiritual birth and mine as well; and my life's work, as a result, has been simply to grow up in what I received, thus the reading, the retreats, the wandering and continued conversation. I have yet to reach an interior adulthood, and the start I have made is surely shaky and fragmentary.

But I have come far enough to see, among other things, that the absence of the esoteric in the European mind, the absence of an essential spirit of *ta'wil*, meant colonial treaty-makers, setting the foundation of this country, were unable to grasp the process they engaged in was essentially a ceremonial event, as it was understood to be on the Indigenous side (see Treaty 7 Elders and Tribal Council 13–15, 323–25 and McAdam 22–25, 64–65). This cultural inability to understand the centrality of the ceremonial and the esoteric, an inattention that formed the world all around me as I grew up, has made the treaty process on the European side a source of injustice, horror and corruption.

Ocean Man in the Lost Horse Hills remains today south of the vanishing town of Corning. All traces of my grandfather's failed homestead and the vacation village at Gooseberry Lake disappeared long ago. The things of my mother's early life are lost, as is, so it seems, the possibility of a conversation about and from within a lived metaphysics, the second world, between First Nations and my people, a possibility squandered, erased by ignorance and villainy. But traces, scraps of the *mundi imaginalis*, remain—the world glimmers and does not refuse the spiritual hermeneutic of an Ibn 'Arabian, Corbanian *ta'wil*. Such a degree of chthonic mercy could transfigure a life.

HAPPY INCOMPETENCIES, THE SELF'S OTHER ROUTES

> Others have enough and more
> I am alone and left out
> I have the mind of a fool.
> Confused, confused. (*Tao Te Ching* 20)

Introduction

Both Augustine of Hippo and Ibn 'Arabi are philosophers who come to bewilderment. In his *Confessions*, Augustine exclaims, "*Nec ego ipse capio totum*" (Myself I do not grasp all that I am) (X.8.15). His puzzlement is not passing or incidental, but substantial; it is also spatial: where could this self be located that I cannot know? So estranged, unfathomable, could this unknowable element lie outside of the mind, or constitute a part of the mind beyond sense? "This question moves me to astonishment," he remarks. Sublimity and its tremor are closer to hand than Romanticism's mountains and seas: it sits within what I believe myself to be. Ibn 'Arabi's foundational brush with ungraspability seems to be placed radically elsewhere, with divinity. But this is a surface observation, since "Whithersoever you turn, there is the Face of God" (Quran 2:105). Unreachability is a ubiquitous trait.

   Both philosophers read their confusion not as failure, inquiry's collapse, but as a fruitful sign one is on the right path. Their confusions, piercing frustrations, while central, make neither an obscurantist. There appears a minute opening. The self and divinity and, one would think, the world, remain approachable by gesture but not understanding. The gestures allowing psychological, as well as theological, access are unusual noetic practices, leading to unexpectedly acute cosmologies and hyperbolic introspection. There are four such routes I can think of associated to greater and lesser degrees with the two thinkers—confession, exegesis, the visionary recital and the projects of eros.

   I wish to put Augustine and Ibn 'Arabi in a colloquy that will be inevitably oblique, and, in doing this, I propose to place confession, *ta'wil*, erotic practice and the dramaturgy of the visionary recital in relation to ontology and ascesis. While I am chiefly preoccupied by *Confessions* and *Fusus al'hikam* and the *Futuhat* of Ibn 'Arabi, as well as the interior theatres of Avicenna, I periodically will take as guides Jean-Luc Marion (*In the Self's Place: The Approach of St. Augustine*[1]), William Chittick (*The Sufi Path of Knowledge: Ibn al-'Arabi's Metaphysics of Imagination*) and Henry Corbin on the spiritual narratives of Avicenna (*Avicenna and the Visionary Recital*), though my conclusions are rarely precisely theirs.

✳ *Confessions* strikes many as a disorderly book. One reason for its spread, its many shifts, is the speeding disquiet Augustine's puzzlement builds in him. The force of this disquiet is also what makes his book conversational address; were he less panic-afflicted, his book would possess more system but less human heat and would not have the form of exchange. There are two loci of confusion-as-proximate-truth in the work, God and the self. "*Nunc autem mihi aliud amo quam Deum et animam quorum neutram scio*" (But in fact I love only God and the soul, about both of which I know nothing) (Augustine, *Soliloquia* 1,2,7, qtd. in Marion 56). But the chief conundrum, at least the most intimate in experience of the power of exigence, is the soul, where "*factus eram mihi magna quaestio*" (I have become a great question to myself) (Augustine, *Confessions* IV.4.9), a difficult country.

In Ibn 'Arabi's psychology, also a kataphatic theology, every name, or essential passion of each being, is necessarily and doubly partial. The name, of course, is idiosyncratic, but also the image it triggers in imagination fails to completely render the being it identifies. Here the block to comprehension is not the wilderness of infinity but a radical finitude, an unfathomable, feral specificity. Further, in Ibn 'Arabi, the deepest "I" is not me and is barred against full approach. But one's name, that numinous flare, unveiling, does speak the truth, the truest truth.

✳ Jean-Luc Marion deplores an Augustinian scholarship, even method of translation, which refuses to abandon the present and its allegiances (Marion 35, 51), that is in fact an ideological recruitment of the dead to serve contemporary preoccupations. How odd to make the past a colony of the present. It is difficult to imagine anything that is quite as intellectually, psychagogically destructive. Augustine must be allowed his own enthusiasms in his own terms. This tolerance, difficult to achieve, involving a brief, blind renunciation of the moment's fiercely honoured orthopraxis, will make him strange, odd-angled, elusive to us, but perhaps the source of welcome, unseating novelty.

Marion sees "*Magnus es, Domine, et laudibilis valde*" (Augustine, *Confessions* I.1; Ps. 47:2), Augustine's core insight, as an analytic remark: you are the deity, consequently great; it then makes perfect sense to praise you. To not praise is to fall into irrationality. To speak truly, to confess, is fundamentally to praise. These words of praise pre-exist in rough form, are speech given to him by divinity: imagination, language and impulse to praise are for him divine artifacts. To speak, considering only the personal, always constrained by the limits of the personal, is assent to this propulsive, pre-existent linguistic force, which seeks itself. "*Quaeram te, Domine, invocans te et invocem te credens in te: praedicatus enim es mihi. Invocat te, Domine, fides mea, quam dedisti mihi, quam inspirasti mihi per humanitatem Filii tui, perministerium praedicatoris tui*" (Let me seek you, Lord, by invoking you, and let me invoke you by believing in you. You addressed me in advance. Let my faith invoke you, Lord, my faith that you gave me, that you inspired in me through the attention of your Son, through the ministry of he who spoke in advance) (Augustine, *Confessions* I.I.13).

Augustine, citing Genesis 3:17–19, remarks, "I have become for myself a soil which is a cause of difficulty (*terra difficultatus*) and much sweat" (*Confessions* X.16.25). He, however, continues to press toward the faint possibility of a route through this impasse. "For our present inquiry is not 'to examine the zones of heaven,' nor are we measuring the distance of the stars, or the balancing of the earth. It is I who remember, I who am mind. It is hardly surprising if what I am not is distant to me. But what is nearer to me than me?" (X.16.25).

The quotation from Genesis refers to the human state after exile from Eden: in its current condition, the self is placed at a distance from itself. As a result, the seemingly central self is hard, modestly productive labour for the self; the opacity of this self, its resistance to analysis, is a sign of this lostness. Nevertheless this difficult, unyielding presence is the sole source of one's particular interior nourishment. This strikes Augustine as a perpetual misery and an unnatural situation: one must address it but with absolutely no conventional hope of correcting it.

Descartes, another philosopher building the edifice of his thought upon the conundrum of the self, in time, is able to assure the ego of its unshakeable existence (*cogito, sum*) and later determines its essence (*res cogitans*). Augustine offers neither clarity and affirms the philosophical truth of this non-performance. At an atomic level, Augustine's approach is a-metaphysical, a-ontological, yet it offers a fresh grounding to both metaphysics and ontology at the close of antiquity.

✴ Ibn 'Arabi refuses the possibility of a certain sort of clarity, that coming from reflective understanding, since this power is reductive in metaphysical and contemplative matters. Here only "opening" (*futuh*)—tasting, unveiling, witnessing divine self-disclosures, inrush—approaches truth. One must prepare for this experience, assemble the conditions permitting its appearance, which arises in retreat, by submitting oneself to "initiatic discipline, that is you must have purified your character, renounced carelessness, and made yourself capable of enduring whatever does you wrong" prior to entering seclusion, he says in his *Risalat al-anwar* (qtd. in Addas 35). A basis for such noetic experiences is prepared outside formal understanding; this pre-knowing, rehearsal for cognitive performance, is itself a series of crucial noological acts.

His own tasting was sudden—"I went into retreat before dawn and received illumination before the sun rose" (Addas 36). Later comprehension followed through the morning. His life became an unpacking of these initial matutinal moments. His hermeneutics during this extended construal was a savouring. "The prophets and the friends among the Folk of Allah have no knowledge of God derived from reflection. God has purified them of that. Rather they possess the 'opening of unveiling' through the Real" (*Futuhat al-makkiyya* III.116.23, qtd. in Chittick xii).

During his pilgrimage to Mecca in 1202, a second series of uncon-cealments occurred to him, which made up the substance of his central book, *The Meccan Openings*. The *Futuhat* follows its author's injunction to "only write what is given by unveiling and dictated by God" (*Futuhat al-makkiyya* II.432.8, qtd. in Chittick xv). Tasting, he concludes, is equivalent to seeing things as they are, both creatures and creator, though this understanding, while true, is partial—thus this knowledge is in fact experienced as a particular sort of bewilderment, an abduction of sense.

This state is in part the result of the fact that all that exists is God's imagination, not the raw material of our own, which itself is the dream of another. Our close sense of self is dogged by the brute fact of an unassimilable strangeness, only part of which is the *haecceities* of things' non-exhaustive divine secrets manifesting accurately as shifting singularities. Divine imagination is also the self's secret, gradually uncoverable over a lifetime until one is all of light's flavours. Moral failure is ontological failure with effects beyond the personal, since all existence is an isthmus between Reality and nothingness. "There is nothing in existence but *barzakhs* [mediating links]…and existence has no edges" (*Futuhat al-makkiyya* III.156.7, qtd. in Chittick 14). This claim by Ibn 'Arabi is not philosophical but mystical theological and ascetical: it sets in place a form of life.

✴ This being the case—that there is only *barzakh*—means that the only thing a person can claim for herself is non-existence. This appropriation is false at a coarse ontological level—of course I exist; there is not a gram of Cartesian skepticism in Ibn 'Arabi—but true at a deep existential point: there is no elemental me that is utterly mine. But rather than extirpating the individual, this annihilating insight into one's actual absence is the onset of radical individuation; the startling band of self I discover in it is a not-me manifesting itself as an unanticipatable, unchosen, but appropriable, numinous name, another light entirely. If one immerses herself in this state, one experiences the individuation of other things—grass, lakes, sandhill cranes—as a consequence, seeing them as they are, other names, and as they are limned in transcendent imagination. Such a self confers an imaginal *haecceitas* on particular things or confesses its indisputable presence, and *haecceitas* is only an imaginal category from an imagination that creates. Thus, a sustaining world appears. The truest realism is theophanic.

How does one find a way to the bewildering Ibn 'Arabian self, the estranging Augustinian self? Introspection courting coherence, psychological analysis, are blunt tools at best; they are almost always truncating. What really is this peculiar self-knowing we have been pursuing? It would be wrong to see it as understanding as one grasps the structure of an institution or an atom: it is a savouring, an annihilation, a praising. The projects of thought that permit mental acts like these include the four ways already mentioned. I will treat in detail only three in what follows, since I believe the noesis of confession has already been sufficiently probed. Erotic endeavour, the labour of pitched interpretation and the enactment of the visionary recital remain to be considered.

The Practice of Eros
Augustine finds himself exiled from himself even in his self-intimacy. He cannot know his essence; it appears not to belong to him, even while being him; he does not confect the image of the life that is his; he knows the self as a "*quaestio mihi*" (a question to myself). This inexplicability is not the result of an absence of interior acuity. It is the result of the intense presence of such insight; the deeper one's introspective probe, the more pressing the ignorance—"*Grande profundum est ipse homo*" (Man is a vast abyss to himself) (Augustine, *Confessions* IV.14.22).

In this helpless state, yet still quickened by appetite—only the soul, along with God, commands interest—the person must somehow go beyond the mind and the self found in memory (X.7.26). He turns to the appetite that acts when memory has performed to its limit: "*Transibo et istam vim meum quae memoria vocatur*" (I will travel across even this power, which is called memory) (X.17.26). I will transcend my own vitality and introspective reach. What is the power that does this? It is the desire for beatitude, a desire for a thing unseen, unknown, contested. This desire—all wish to be truly happy—precedes choosing, wanting, acting; it visits one from a place before the electing conscious self, coming, it seems, from an a priori, unexperiencable place.

Eudaemonic craving is extra-personal or at least outside the mind I view as my own, the achievement of my own construal and choosing, yet it ceaselessly recruits this mind. It alone authentically complements the self, but jarringly by unselfing it through a de-identifying elongation. Indeed it is oddly the prime source of identity, this faintly alien, not-wholly-confiding prior force. One knows one's core desire not notionally, not in willing, but in an almost physical bowing before exigence, cannot know it since it precedes knowing and utterly moves and shapes it. It acts either through assent or fragmentarily through suppression, and in both situations one notes the back of this advancing-through power and reads it from behind as it moves ahead.

The formative push is a further strangeness and unselfing of the self. "Certainly we have the desire for it [the happy life], but how I do not know," Augustine remarks (X.20.29). Desire precedes knowledge but also desire itself and pedagogically exercises it, shaping it roughly for the *beata vita*, the blessed life. It exists in memory but prior to recollection, inaccessible to recollection; "the happy life is found in memory and is recognized when the names are uttered" (X.21.31), a hyper-intimacy, recessed but articulate within the intimate self. The mind, ingenuity, desire's pre-knowing, its extraordinary, aspirational gestalt sensitivity, its sad, indefatigable energy steps forth when one has her desire uttered through her. In Ibn 'Arabi, this Augustinian practice of eros—a self-listening and a mimesis of the shadowy a priori yearning self—is the ascesis of becoming one's name-cargoing passion.

The practice of desire involves the development of an erotic connoisseurship—"*haec est vera delectio ut inhaerentes veritati juste vivamus*" (This is the delectation: to live justly while clinging to the truth) (Augustine, *De Trinitate* VIII.7.10, in Marion 94). Yet before this recognition, what one essentially wishes to have or be eludes full disclosure, but is vivid in the discriminating nose of toned desire. The self is home when it is beyond what it knows and in a place desire, once its discernment forms, will confirm.

"But you were more interior to me than what is inward in me and higher than the highest element in me" (Augustine, *Confessions*

III.6.11). What is sought is an intimacy that exceeds the cogito, exceeds psychoanalysis, the credo, intentionality, that precedes and exceeds the "I" and supplants, enlarges and fulfills the "I," revealing itself as the self's secret identity and wealth: *"Ibi mihi et ipse occuro meque recolo"* (Here I meet myself and recall what I am) (X.8.14). The deeply recessed, acutely apt inner person is found with the one who inhabits him, as Augustine observes in *De vera religione* (xxxix.72, in Marion 346), though the desire carrying him to this discovery declares no destination, since it knows no fixed object, even if it is fitted to one. It tracks ahead, investigating scents; it confirms in delectation the approaching arrival at the always wanted term.

Ta'wil and *'Ayan*

In the *Futuhat*, Ibn 'Arabi proposes a taxonomy of intoxication: natural intoxication exists in "the delight…found by souls through the in-rush of wishes," rational intoxication in attachment to proofs, and the final intoxication that is with God. In the delirium of this last state, one cries "increase my bewilderment in Thee, for the drunkard is bewildered" (*Futuhat* II.544.16, qtd. in Chittick 199) and on the path aimed at a sifting of the self. This wish for bewilderment as a philosophical platform is not a sly move toward some sort of vitalism, in which one is some elemental force's sleepwalker, but a station on the way to full self-knowledge and a trued choosing.

There is the wish, abandoned post-Kant, to know the world as it is in its swarming individualities, to uncover the precise names and unparalleled essences—"[t]o see their identities—or, if you so wish, to see His Identity" (Ibn 'Arabi 3). The uncovering of names comes to *'ayan*, variously rendered as "spring," "eye," "source," "quintessence," "essence," the ultimate identity of a thing. Specificity, particularization, thus interpretation, is the externalization of a thing or person's identity, which is the same as the object in the divine mind, so its release is a revelation of an aspect of divine relating, a form of God's manifestable mind. What is this way of seeing identity that is a piercing realism, as well as kataphatic theology? "In general, 'identity' carries with it a certain dimension of inwardness and transcendence, and it enshrines the meanings of essence, sameness and individual nature, all of which are inherent in Ibn 'Arabi's use of 'ayan,'" says Caner K. Dagli in his

"Translator's Introduction" to the *Fusus* (xix). So this seeing is a fusion of individuality and divine nature, the specific and the universal.

Each being is a name of the identity of the universal—so sociability is an epistemological and philosophic erotic duty. But, again, what is the seeing that reveals identity? It has no real contemporary label because its operation has been hounded beyond the margins imagined for human capacity. "[I]t requires a great deal of courage today," remarks Henry Corbin, "to invalidate, in the name of a spiritual interpretation, conclusions drawn from archeological and historical evidence…" (*History* 3). In regard to this banished hermeneutic, Corbin insists on a sharp division between allegory, which is harmless, and symbol encountered in a spiritual exegesis of texts and events, which is revolutionary. The cognitive mechanism of the spiritual hermeneutic involves ekstasis, the acme of sociability—a radical empathy, dwelling in the other and in my deepest self. Here is Dagli once more: "A self or subject is by definition not an object, which is to say the very self of a being is the mystery known only to it and to God. To truly know my selfhood as such, one would have to *be* me, which demonstrates the immutability of the self" (Ibn 'Arabi xxv).

In the Perfect Man, or "man of light" in Henry Corbin's phrase, every possible identity inheres; he is a "small world" and an *imago Dei*. "It can be seen that man's most inward and transcendent reality is itself the principle by which all things in the world, including himself, are made to exist," says Dagli (xxx). The sociability, the spread of this knowledge, is achieved most acutely by the introspection of the "perfect man." The seeing of identity here, discernment, *alētheia*, is the result of interior formation and the dilation this brings to physical and contemplative senses. One knows the sacrality of the world in discerning and realizing one's name in one's behaviour and life. The name, unusual substance, appears out of a state of personal non-being.

Through *'ayan* you see the essence of particular things, but you see them firstly and most profoundly in yourself: and this is a way to self. This passage to self is simultaneously an asymptotic passage to divinity and the world. "All of the Names, which are divine forms, are manifest in the make-up of man, and the function of encompassment and synthesis is achieved through his existence" (Ibn 'Arabi 6). To go further: as the *Fusus al-hikam* claims, it is more accurate to describe seeing things as they are as divinity seeing itself and existentiating the thisness of things in this single act. Our crispest subjectivity then is a reduced form of divine subjectivity, divine self-recognition, our higher mind knowing itself in extension to everything as we come to be.

"The Real willed, glorified be He, in virtue of His Beautiful Names, which are innumerable, to see their identities…" (Ibn 'Arabi 3); this act is the curiosity that is creation. It is not a creative act but an extra-personal introspective one, a self-solicitude and branching generosity, since these identities, *haecceities*, names, are "His Identity." This inquiry is satisfied only if the yearning is transplanted "in a comprehensive being," which, Dagli tells us, is the existent world; by this act of sharing an extraordinary subjectivity in the recognition of the natures of things, "His Mystery is manifest to Himself through it" (Ibn 'Arabi 3). What is not reported in this act of seeing is divine essence, which is beyond all powers of recognition. So here we have a complex self indeed, divine curiosity, self-recognition, the manifestation of massive individuality and the end of divine noetic yearning.

An illustration of this cognitive performance at work is the triple-seeing in Christ's recognition of Nathaniel in the Gospel of John (1:47–51), where an augmented self appears within a vastly more ample realism.

> When Jesus saw Nathaniel coming toward him, he said of him, "Here is truly an Israelite in whom there is no deceit!" Nathaniel asked him, "Where did you get to know me?" Jesus answered, "I saw you under the fig tree before Philip called you." Nathaniel replied, "Rabbi, you are the Son of God. You are the King of Israel." Jesus answered, "Do you believe because I told you I saw you under the fig tree? You will see greater things than these." And Jesus said to Nathaniel, "Very truly, I tell you, you will see heaven opened and the angels of God ascending and descending upon the Son of Man."

Such realism is a form of *adab* (courtesy) toward being, courageous, merciful in its spread.

The Visionary Recital

The core of Ibn 'Arabi's epistemology is that Reality's "self-disclosure" is "the lights of unseen things that are unveiled to hearts" (*Futuhat* II.485.20, qtd. in Chittick 216). The mirroring lobes of unveiling are ontological and cognitional: when he sees things as they are, he finds himself and his place in being. He comes to this seeing through listening and "the spiritual concert," both aspects of *sama'*, "audition," and through ecstasy. In the view of the Sufis, "*wujud* [Being and the cognition of unveiling] is finding the Real in ecstasy," an "unexpected occurrence" (*Futuhat* II 538.1, qtd. in Chittick 212). The view of being achieved in this state is idiosyncratic yet universal.

> The finding of the Real in ecstasy is diverse among the finders because of the property of the divine names and the engendered preparedness. Each breath of engendered existence possesses a preparedness not possessed by any other breath…Hence the finding of the Real in ecstasy takes place in keeping with the divine name that watches over him, and the divine names go back to the Self of the Real… (*Futuhat* II.538.1.21, qtd. in Chittick 213)

If an efficient block to such ontological, self-introspective illumination is "rational intoxication," the mesmerization of the proofs, how might this be breached? We have already investigated two means: eros's genius and the fully dilated, panoptic hermeneutic. But these operate at deeper levels of initiation, once one is fully turned to the task. A device at play at the threshold is the visionary recital, where, as Henry Corbin remarks in his *Avicenna and the Visionary Recital*, "dramaturgy" (125) is substituted for cosmology, thereby guaranteeing "the genuineness of the universe; it is veritably the place of a personally lived adventure."

The equivalent of Avicennan psychagogic performance in Augustine is the theatre of conversation, his drama provoking transformation: it induces its own intoxication, a space of maximal suggestion in interior theatre, the thrall of performance. He quotes from the Psalmist, "*Quoniam cogitatio hominis confitebitar tibi et reliquiae cogitationis sollemnia celebrabunt tibi*" (For man's thought will confess you and the remnant of his thought will praise you) (Augustine, *Commentaries on the Psalms* 75,14). His is a cognitional theory premised on an exigent, innate dramaturgy of speech to another, where thinking, under the eudaemonic urge, with nowhere else to go, becomes an inevitable performance of inner seeing and praise.

Avicennan visionary recital has a similar ascesis and terminus, but a different means. In his spiritual romances, the powers of character development, narrative incident and culminating heroic acts, an initiatory choreography, deflect a rationalist exegesis of subjective states and force a daemonic entry into writing and thinking.

The second in Avicenna's cycle of performances is *The Recital of the Bird*. Its prologue, a will-assisted ascetical prelude to enrich participation in the drama, stage directions forming attention, urges a disposition mingling audacious intimacy with hermeticism. "Brothers of Truth!...Meet together, and let each raise before his brother the veil that hides the depths of his heart...Brothers of Truth! Retire as the hedgehog retires...it falls to your hidden being to appear, while it falls to your apparent being to disappear" (Corbin, *Avicenna* 186–87). So the theatre is the lit stage of interiority, the narrow seats before it. Softly move in this place—"[w]alk like an ant" (187). One is prepared for Corbinian *ta'wil*, the acute, spiritual hermeneutic, convivial, vigorous, daring. "Be ever in flight...If you have no wings, get them by sleight...Be like the salamander that lets itself be wrapped in flame, at ease and confident" (187). Now shaped by these admonitions, the reader (auditor) is moulded for the recital, having undergone an ascesis preceding reading, reading's essential antechamber, without which there is no reading.

The recital's bare narrative concerns the situation of snared birds, who have been lured into nets by hunters' delightful calls. The speaker is one of these birds. After a struggle to free himself, he becomes resigned and slumps in the net, only to be rescued by wild birds. He joins them in a flight over seven mountains and takes rest on the summit of the eighth, a place of "green gardens, beautiful fruit trees, charming pavilions" (190). This moment is a training of taste. Tempted to stay, he elects to press on and reaches the ninth mountain, equal to the Ninth Heaven, where the bird-souls meet their "celestial family"— "they treated us with such charm, delicacy, and affability that nothing created could describe it or make it comprehensible," the yearned-for but unfound, comprehensive intimacy of all searches for home. They advance farther in this place to the oratory of the king, where all his beauty shows, and "our hearts hung on it and were seized with a stupor" (191).

The Recital of the Bird resembles philosophical formational narratives in Plato, the Allegory of the Cave, Socrates's account of his education at the hands of Diotima, the Atlantis myth and the flood of the ideal

Athens in the *Timaeus*, among others. The point of these stories is the recovery of the actual self, a new, unanticipatable being. Socrates's interlocutors unfailingly flinch from this hidden, larger, unguessed, discomfiting self. In the recital's epilogue, Avicenna predicts the story's fate: the extreme speech will be rejected as heterodoxy, insanity or be greeted by cries from the heart of metaphor blindness—"How should a man fly? How should a bird fall to speaking?" (Corbin, *Avicenna* 192). These reactions will protect the psychagogic power of the tale in occultation and prepare for its hermetic infiltration by those able to manage this.

✳ The true self is the one you stand before confused, uninvited, unassimilable. It is *"grande profundum,"* larger, more, *"terra difficultatus,"* and the one thing, aside from the ground of being, the introspective mind loves, and through which it plants itself, true, in the world. Before the catastrophe of climate change, such exercises as those that provoke these encounters will be seen as unforgivable indulgences. But this charge is part of the multifaceted sadness of this time. The route to a durable, just, pan-ontological politics runs through a corrected epistemology, a retrieved, heterogeneous, ample subjectivity; and scanning the path to be taken, one is struck—heart-struck—by how much has been culturally lost, so that this retrieval of maieutics from the deep past seems barely likely. Yet these acts and their knowings *must* be re-gathered, re-said, re-enacted even as the seas rise and species vanish. This retrieval is one way to cope with and, with far more luck than we can justly hope for, lessen these losses.

Note

1. All quotations from *Confessions* owe much to Marion's translations.

POVERTY AND THE DOOM OF ACEDIA

JEAN LECLERCQ, OSB, in his preface to a book that resurfaces the thought of Evagrius Ponticus (345–399), writes about the pursuit of, the possibility of, "peace: that calm, that security, that repose, that Sabbath, that leisure, that reality so rich that it cannot be circumscribed by any words" (Leclercq xii), which is the goal of the interior life. This peace, says Leclercq, is what Evagrius, monk, acute psychagogic diagnostician, meant by *hesychia*, a solitude of spirit. The search for *hesychia* stood at the heart of contemplative practice in fourth-century Egyptian monasticism, where Evagrius himself played so central a role.

Language creates, by some trick of vivification, objects, interior states, lives; it helps them breathe, be and participate with other lives, states and objects. Language quickens worlds and sustains them. It makes certain emotions, psychic postures, available to sense, occupiable. I say in W̱SÁNEĆ territory, on the west coast of Canada, ĆEN IŁĆ, and the reality, bristling presence, the animation, the generosity, mercy of the Garry oak that rises to the top of the cliff on the mountain SN̲,AK̲E and lays its highest branches on a small plateau becomes not only more evident but also tends toward me, finding a way to me I could not have guessed. I say apokatastasis, the restoration of all things, the resetting of the ontological bone, and an extraordinary political and personal attachment is stirred and stretches a life that before had seemed bounded by stern walls. This vivification of states and objects by language is related to courtesy and daring on the part of the naming subject, as well as covered plenitude on the side of the named. Autocrats learn that a simple way to subjugate a people is to flatten their language. Catastrophe is deepened when language is impoverished. The expansion of language through the retrieval of ancient wordscapes of the inner life and silenced-by-colonialism Indigenous languages is healing, en-homing, life-preserving and subversive. A fuller life can come from both these sources.

We descend now into another linguistic vein. It has been worked, well worked, in the past, but many of its central words seldom now find themselves in a mouth or issuing from a pen, though what these words name unceasingly boils in hearts or are mutely longed for. Its strangeness notwithstanding, this language, these languages, can help us now as nothing else can.

✶ Eight dispositions deflect someone given to interior attentiveness from the peace into which Evagrius wished to release his readers. These misdirecting states, by his reckoning, are gluttony, impurity, avarice, sadness, anger, acedia, vainglory and pride, all corruptions of eros. "It is not in our power," Evagrius wrote in his *Praktikos*, "to determine whether we are disturbed by these thoughts"—they inevitably settle in all but with unusual ferocity, in the blaze of combat in fact, in those engaged in contemplation—"but it is up to us to decide if they linger within us or not and whether or not they are to stir up our passions" (par. 6). He described antidotes to these unbidden inclinations, some unexpected: "Turbid anger is calmed by the singing of Psalms, by patience and almsgiving" (par. 15). Never let "the sun go down on your anger" for, then, you may find yourself jolted in the night by vivid recollections of the initiating offence in 2:30 a.m. self-to-self colloquia and the anger, unlanced, soon becomes something more threatening.

"Then there comes a time when it persists longer, is transformed into indignation…This is succeeded by a general debility of the body…" (par. 11) and finally deep-seated anger turns to "desertion." "[N]othing is more disposed to rend the spirit inclined to desertion than troubled irascibility" (par. 21). What he means by "desertion" is the unwitting abandonment of one's entire source of meaning.

Everything caves when meaning evaporates. Some of these eight tendencies kick the door in, lust, anger and vainglory, say, while others, like sadness and a particular dolour called acedia, slither in through cracks. "Sadness tends to come up at times because of the deprivation of one's desires. On other occasions it accompanies anger" (par. 10). It provokes a caustic nostalgia for times past, a longing for those people and circumstances you have lost: you dwell on the injustice of the passage of time; the lost moments are now "drenched" in sorrow in memory and "the miserable soul is now shriveled up in her humiliation that she poured herself out upon these thoughts of hers" (par. 10).

The most destructive visitor is acedia, and it comes not at introspective dusk, or in the middle of the night when you are most vulnerable, indignation's best hunting grounds, but in the vigour of the day from around noon until late afternoon. It has many features—anxiety; disgust with one's situation; disgust for oneself; an enervating boredom; a rejection of the beauty of all the options; the conviction that one would be treated better, his merit better seen, elsewhere. Under the spell of deep acedia, one's present life feels unbearably light, under done, repulsive, not durable in the least; weariness sets in, then is replaced by frantic, ineffectual, re-inventional activity as you try to re-energize or re-situate yourself. The febrile do-overs trigger deep interior tremble, parching fret. You skip about in mind and heart; you are afflicted by a feeling of disdain or contempt for those you live with, the unacceptable milieu within which you must function. All this roil of emotion, or significant aspects of it, comes "rushing in" like a fever, says John Cassian (*Institutes* 219); you are "enfeebled" within, a "slackened oar" as Dante observes in the *Purgatorio* (XVII.l.87). This state is the questing heart's greatest obstacle, its knot the hardest to untie. To stand in it can feel like an exceeding sensitivity to the reality of one's state of affairs, a sharp, clear awareness of what justice in one's life demands, but this is a false depiction of where you now find yourself.

✳ If our world, the world of capitalism and the untethered, assertive, autonomous self, as Kathleen Norris argues in her *Acedia and Me: A Marriage, Monks and a Writer's Life*, is acedia writ large and monetized, this interior condition I have just described is unavoidable and its fever rages free, a cataclysmic norm, and the acts it crafts—various strainings to tear down and rebuild, various precarities of spirit—and their debris pile around us.

Donald Grayson, in his book *Thomas Merton and the Noonday Demon* observes that monks in this state "were separated from their true or deep selves, from their monastic brothers and from God" (25–26), contracted into private pain and shedding longing in a shard of the self, a condition that Grayson, echoing Paul Tillich, calls sin as separation. Depression, for which acedia can be mistaken, Grayson reminds us, is a medical condition, while acedia itself is a sin, an existential error, a mis-seeing, now even more difficult to make out, given its ubiquity, and difficult to free yourself from. Yet defeat of acedia, says Evagrius, stands next to the experience of *hesychia* (par. 12).

In late antiquity, in Evagrian psychagogy, in desert ascetical theology in general, acedia was consistently associated with sloth, manifest, in part, as self-pity and hatred of manual labour (par. 12). This appraisal is likely to strike us today as unjust. If acedia now is diagnosed as a type of depression, as I suspect it often is, and is indistinguishable from it for most of us, describing it as a form of laziness feels like blaming the victim and refusing to see the actual, objective nature of the sufferer's affliction. Besides, those caught in acedia don't usually seem torpid but hyper, chokingly industrious, febrile, in the matter of fantasized self-definition. I think acedia's primary sloth is real and interior, a slackness in the imagination of discernment, an unwillingness to introspectively probe honestly deeper. It is, in its earliest moments, a precipitous and tentative noetic and emotional desertion of where one stands, a refusal to be patient and "[b]ravely take all that the demon brings upon you" (par. 28), investigating and knowing it, digging into its misidentifications, as Evagrius recommends we do. Acting out of unexamined, unchallenged acedia produces not freedom and light, he warns, but instead fosters a life of awkwardness, cowardice and fear (par. 28).

✶ Stability seems to be the first step in acedia's treatment. "The time of temptation is not the time to leave one's cell, devising plausible pretexts" (par. 28), advises Evagrius. Be patient and take on the full load of what the demon brings. The counsel seems merciless, wrong-headed, but the condition needs to be sifted and named.

Acedia, the engine room mood of consumer life, is the source of its perpetual uprootings, strategically tentative, oddly slanted selves and alliances; it is the place of protean, persistent longing for golden elsewheres, even more gleaming celebrities, liberalism's principled unsettledness and interior anarchism, issuing in increasingly polished or perfectly positioned, or apparently more deeply appropriate, self-stagings. Rather than acquiescing to the flow of the compulsions, try to stand calmly among them, comprehending the nature of the threat and the force of the mid-channel turbulence.

✶ "The other demons," anger, vainglory, gluttony and the rest, "are like the rising or setting sun in that they are found in only a part of the soul. The noonday demon, however, is accustomed to embrace the entire soul and oppress the spirit" (par. 26). The acedic soul may feel it has no rope to hold in imagination aside from the chief symptom of its problems, the plethora of schemes for escape, re-invention, impossible return or release that build the fever. Evagrius, though cautious of this aspirational fret, this worked-in dissatisfaction, however, does *not* oppose change; rather, he champions *conversatio morum*, a commitment to perpetual alteration, perpetual deepening, his response to what Gregory of Nyssa, an early influence, saw as the infinity of virtue. "[O]ne limit of perfection," Gregory noted in his *Life of Moses* (Prologue, 5), "is that there is no limit." To correctly grasp the ways of this continuous, essential change Evagrius counsels discernment.

The former archdeacon was renowned in the desert not so much for his books, and not at all for his eloquence, but eventually for his discretion, his gifts of "knowledge and wisdom and the discernment of spirits," as Palladius wrote of him in *The Lausaic History* (38:10). This acuity concerning the interior life came to Evagrius slowly as he moved from the highest ecclesiastical offices in Constantinople, where he served as a homilist, philosopher and imperial counsellor, to a quite different life in the crude dwellings of hermits in the desert of Nitria. There, under the tutelage of such monks as Macarius of

Scete, who himself followed the example of Anthony, the first of the monks, Evagrius came to acquire trustworthy, sharp-eyed judgement, a merciful ability to make out the idiosyncratic shapes of salvation. Evagrius learned the skill of diaresis, of distinguishing between apparently similar things or states, or had it grow in him, what makes for the spiritual guide, through an ascesis of humility, silence, dwelling on his own struggles and the meanings of his theophanic moments and by absorbing the model of his teachers, living and dead. The power of discernment also requires a modicum of calm, a space of *statio*, a standing still, in order to grow. The achievement of such a flickering stillness is part of the psychagogic cure of acedia, an initial stage, an interior *stabilitas*, as one is lifted, for a moment, from acedia's clamour.

✳ Anthony, "father of the monks," founder of the eremitic tradition, trail-blazer for those drawn to a life shaped by contemplative attentiveness, early in his time in the desert, was devoured by acedia, its streaking thoughts, its incessant lightenings. Seeking a way out with great intensity, he one day had a vision: "…when he got up to go out [from his hermitage] Anthony saw a man like himself sitting at his work, getting up from his work to pray, then sitting down again and platting a rope, then getting up again to pray" (Ward 2). The man, Anthony surmised, was an angel sent to perform him in order "to correct and reassure him." "He heard the angel saying to him, 'Do this and you will be saved'" (Ward 2). This apothegm is placed at the head of Benedicta Ward's alphabetical collection of patristic sayings, suggesting the deadliness of acedia's threat. This alternation between work and prayer given to Anthony became the sustaining form of monastic life for 1,700 years, the essence of Western contemplative practice.

While I see this oscillation between interior attention and physical, often food-growing labour is a rescuing discipline, I suspect it might not be quite enough to disarm the menace of the toothed boredom of the late afternoon, which has the power to draw you into its tearing nothingness. But it may be sufficient to create an unstable calm sufficient for *stabilitas* in which a discerning mind may form. From there, a deeper thinking and a more complete healing might be attempted.

✳ As we live into climate change, and as I consider a retirement from more than thirty years of teaching, I sense different voids forming around me. There is the unbordered emptiness, the aching, dumbfounded solitude, of ecological mourning, and apart from this but communicating with it, the spreading noon of a green acedia, appearing as frantic activisms or grand despair. I admit I have fallen into the pit of acedia's torpor on occasion, where it's hard to lift a finger, open a mouth, stranded in a zone of unignition, unmeaning, appetitelessness, pulled nowhere. I've tasted, too, acedia's daily blunting frenzy: who am I; how shall I be in this turmoil and defeat? I speed back and forth with various resolutions; my noons fill with itchy longings, engulfing plans, that I can see are futile even as I remain mesmerized by their flash and velocity.

To reduce the febrile strain, I apply Anthony's angelic alteration: I write or read in the morning and chop wood in the yard or work in the community garden in the afternoon. Sometimes the "joy and courage" promised Anthony, rising from the back and forth sway between *ora et labora*, come and sometimes they do not.

I have been reading Bonaventure, a thirteenth-century Franciscan, for the last few years because I have wanted to understand better what he calls the agent intellect, "the journey of the mind to God," that part of thinking where the unexpected, but apt, appears; and also I want to learn more about his notion of imagination, yet another manifestation of the mind advancing from outside the mind, the mind before the mind. Completing his *Itinerarium Mentis in Deum* just before Christmas, freshened by his view of the divine infinity participating in the human person through imagination and osmotic noesis, I turn, as another summer begins, the first COVID-19 summer of self-isolation, the summer I turn seventy, to Bonaventure's *Life of Francis* and read his attempt to grasp the nature of what he called Francis's "ecstatic peace," a state Bonaventure found both mysterious and boundlessly appealing.

✳ Bonaventure's *Life* is less a biography than a gathering of luminous bits of behaviour of a man who, while dead, continued to instruct him, a man of "superhuman affability" (Prologue 3), who practised a "generosity beyond his means" (1.1), going back to his earlier secular life, so that these phenomenological fragments "may not be lost" (Prologue 3; John 6:12). Bonaventure reports Francis fell ill in his early manhood, *divina afflictio*, the perennial numinous prelude, that portal, and in recovery, he happened to meet a knight who was down on his luck. Even then given to the radical gesture, Francis immediately clothed the knight with the garments he himself was wearing. The next night Francis dreamed a splendid palace full of weapons and other military gear and was told that these were for him and those who follow him. It would take him a long time to decode this revelation, but he persisted in puzzling over it, taking this mastication as a discipline, slowly digesting the obfuscated epiphany. He soon, though, turned over his wealth to his disputatious father and, leaving Assisi "in a carefree mood sought out a hidden place of solitude where alone and in silence he could hear the secrets God would convey to him" (Bonaventure, *Life* 2.5). Francis had found his sweetness with the occupation by another mind, in his own permeability and lack.

Francis also came to discover in himself the appetite to obey—he learned this taste—at all times, the angel, which, in certain epistemologies, particular ontologies, Ibn 'Arabi's for instance, or Bonaventure's own, is in fact the active intelligence, or at least speaks on its behalf. Attending a Mass of the Apostles shortly before the founding of his order, he heard the injunction that those sent out to preach, that is, to engage in psychagogic conversations, "should not keep silver or gold or money in their belts, nor should they have a wallet for their journey" (Matt. 10:19), and he believed he had found an unexpected treasure. "This is what I want; this is what I have longed for with all my heart" (Bonaventure, *Life* 3.1), this feeding nothing through which he will sustain himself and through which the unlike would speak.

When Francis's first follower, a man named Bernard, asked him how he should live, Francis suggested that the two consult scripture. The next morning they randomly opened the Gospels in three places: "If you would be perfect...sell all that you have and give to the poor" (Matt. 19:21); "Take nothing on your journey" (Luke 9:3); "If anyone wishes to come after me, let him deny himself and take up his cross and follow me" (Matt. 16:24). The discerning exercise is itself intellectually threadbare and it uncovers a counsel to poverty. Francis recognized immediately that a dawning essence in his unfolding, self-slimming desire had been spied out: "This is our life and our rule," he declared with conviction and with what must have been relief (Bonaventure,

Life 3.3). We have uncovered who we are; this is the actual name of our latent, persistent desire. Poverty, in all its forms, he sensed, was the ideal end-point of his epistemological and erotic reaching and the means by which a true, novel, sustaining identity was to be acquired.

This emptying also creates a land of incalculable epiphany in which one may live in delight, constantly transfixed by eruptive readings of daily events. The small band of the earliest friars made their way to the valley of Spoleto; as night fell, they found themselves in an isolated spot. They realized they had no food with them when "suddenly a man appeared carrying bread in his hand…" (4.1).

Francis's poverty is emotional and epistemological as well as material; in his feelings, thoughts, needs, he steps back from himself in order to find the appropriate disposition, in order to know and be nourished. He occupies the commodious vacuum of lightly entertained anticipation: "Holy poverty,/which was all they had to meet their expenses/made them prompt for obedience…Their very poverty/seemed to them overflowing abundance since, according to the advice of the wise man,/they were content with a minimum/*as if it were much* (Ecclesiasticus 29:30)" (4.7). Sufficiency is tabulated by means of a new psychagogic calculus. The sustenance, the wealth, is already given; it awaits its second birth, as pleasure and vocation, through discernment.

✶ On Christmas Day, 1952, Thomas Merton wrote to the prior of one of the two Camaldoli monasteries in Italy expressing his desire for a life of complete solitude in the hermitages (*eremo*) under Anselmo Giabbani's leadership. He wrote on the sly, without his own abbot being aware of the exchange. "O my father," Merton wrote to the prior. "I have an immense need for that silence, for that *quies* [a mute lodging in God] in which the soul rests, unmoving, in the obscurity of an immense and simple activity which is God himself, communicated to the soul by the ineffable light of the Word and in the embrace of the Holy Spirit." He would find this *quies*, he believed, only in the eremitic life of Camaldoli. He pronounces himself drained by existence in his home monastery, Gethsemani, and disdainful of the "monastic vanity" found in the cenobium, the community of many brother monks, which "suffocates souls," he claims, the cenobitic life that requires "one must submit oneself, at whatever cost, to a kind of liturgical autocracy which, I admit, suits the majority of monks very well, but which cuts off the spiritual breath of those souls whom God calls to simplicity and solitude" (Grayson 70). Merton concludes that he is *"not made* for this" (71).

This exchange with Giabbani, Merton's back-door attempt to remodel his situation, even though he had never visited Camaldoli and, as Grayson convincingly shows, had a skewed romantic sense of life there, continued for several years. The industry and secrecy of this attempted self-definition is an outcome of acedia: I feel myself obliged to take on the mammoth task of articulating what I am and setting in place a structure allowing me to be this. Most means are acceptable in this endeavour. My present circumstance is irredeemably corrupt, and I mar my singular identity by submitting to it. All this activity takes place in a realm of possibility, which allows for its ample expansion, but this distance doesn't offer interior protection to the sufferer of acedia. The fever builds; its damage grows. These machinations separate one affectively from one's fellows and their ability to see with some clarity, or at least not as you see; the busy theatre of constructing these futures in thought lessens the ability to truthfully read oneself, the pattern of one's foundational eros, and also blunts the capacity to taste providentiality. This committed fiddling indulges the habit of privileging a particular moment of judgement, one that defers to the outraged self, and this weakens the talents one wants to free through these new situations.

Thomas Merton, after a decade at Gethsemani, was clearly restless, and there were other golden elsewheres he entertained, the Carthusians, Ernesto Cardinal's new contemplative foundation on the Nicaraguan island of Solentiname, a monastery in Mexico, another monastic possibility in Ecuador, a febrile spread of enticing new lives, but Camaldoli was the preeminent draw. This existential activism, a busyness around rearranging the self, re-positioning for supposed maximal effect, while laudable perhaps from a liberal, *On Liberty* perspective, a mark of the altruism and nobility of aggressive, inventive Millian self-determination, looks quite different from the perspective of a Franciscan love of poverty, which is a sagacious erotic passivity.

✷ Bonaventure, in Book Seven of the *Life of Francis*, continues his description of the early years of the confraternity. The order's cohering ethos was found in its members' delight in the miraculous nature hidden within lack. When the friars asked him at a gathering what virtue does more to make one a "friend of Christ"—note that in Bonaventure's epistemology the active intelligence shares an identity with the Logos— Francis "replied as if opening up the hidden depths of his heart 'Know, brothers, that poverty is the special way to salvation, as the stimulus of humility and the root of perfection, whose fruit is manifold but hidden'" (7.1). A "friendship with Christ" in Bonaventure's cognitional scheme is a discipline of attending to, cleaving to, the visiting intelligence that is Logos received as a result of self-abandonment and nurtured discernment. Francis continued, "This is the Gospel's treasure *hidden in a field* (Matt. 13:44); to buy this, we should sell everything, and in comparison to this we should spurn everything we cannot sell" (Bonaventure, *Life* 7.1). The pattern Francis grasps in Being is the occultation of unaccountable sufficiency, the root of delight, its sequestration from noesis and effort. This wealth, though, is not arcane but simple; it is the wealth of the world awaiting apprehension, the stunning abundance of singular being continuously showing itself.

This providence, an emergent, non-imaginable excess, kept the friars safe, rapt. "They were safe everywhere,/not held back by fear, nor distracted by care;/they lived with untroubled minds/and without any anxiety/looked forward to the morrow/and to finding lodging for the night" (4.7). Poverty appears with *amor fati*, with seeing precise fullness in little, with obedience not only to the friends of one's soul but also to those charged with one's welfare, but without a special care for one: all this insightful apprehension draws down the fever of acedia. And the world comes toward you with risibility and with riches in its arms: discernment's imagination allows you to see this perfect wealth as you strain for perpetual deepening in virtue's endlessness. "Their very poverty/seemed to them overflowing abundance/since, according to the wise man, they were content with a minimum/as if it were much" (Bonaventure, *Life* 4.7; Ecclus. 29:30).

Acedia is the desperate, inflated craving and flight-dreams, the hollowing, freezing boredoms, that can arise around, or take the place of, interior alertness. It is a distracted fussing over a supposed future due to the feverish self in its various inflammations. But if that self is given, donated to the apokatastatic flow of the Second Person of the Trinity, the genius of the active intelligence, Being, that repeatedly novel excess, caught in discernment's nimble mapping, its otherness and elsewhere are perpetually and decently found. The self that acedia assiduously and demonically serves has decamped, and the self as donné, Francis's self, finds nothing that appeals in acedia's dream. Franciscan poverty disables the drowning, the flailing, the grasping of acedia, its repetitive dislodgings.

The soul in the throes of acedia isn't troubled by an inability to donate the self—it does this only too readily in imagination in its rushing pleonexia—but cannot discern, in its rushing, what form of life will give it what it truly wants, its form of *hesychia*. It also lacks courage, which would provide stability, and industry in thinking through its predicament; industry, too, would contribute to inner *stabilitas*. Firmed up by courage and industry, one can bestir or form the genius of the inner palate, the power to make out the completing beckoning just beyond what you know, a sense of what is justly, beautifully, necessarily next.

Acedia's self is rampantly acquisitive, eating imaginary meals, imagining enviable others eating their even better ones, collecting in an incumbering way deeply into faux satiation. Francis's self is unstintingly kenotic, teleological only in a broad sense—it wants to love and fully live—but with nothing to say on the details, the specific steps of this quest, which it leaves to an outside and defamiliarizing genius that comes to it via the passageway of an interior connoisseurship. Francis's poverty is sharply discerning: I give all so that I may receive a larger all, which is the truth of providentiality, the sufficiency of being. Nothing acedia plots can match this, even if it were to hatch a scheme to gain the whole world, traditionally the error of the once beautiful angel Lucifer.

I take up for myself in this time of pandemic disease, land-based sorrow and drivelessness a life formed by a blend of Benedictine, that is desert, insight into labour's theurgy and prayer and Franciscan poverty and freedom. I do not exist in these states, though I pursue them. While poverty is one of the three Benedictine vows, Franciscan poverty seems to me more complete, celebratory, ecstatic, instructional, ingenious, reaching more deeply into intelligence, more expansively hopeful and ambitious and more available to the wildly protean world. In fact, this poverty creates the livable, vivifying world by its self-escaping acts and imagination and is, in this way, a politics, indeed a revolutionary one.

ONTOLOGICAL LONELINESS
AND THE BALM OF METAPHOR

*Talk given at Beijing Normal University, Beijing, China, March 17, 2015,
and at Trinity Western University, Langley, BC, June 2015*

PSYCHOLOGICAL COUNSELLING, psychoanalysis, is a multi-million dollar industry in North America. It takes a variety of forms—Freudian, Jungian, Adlerian and various narrative therapies like Viktor Frankl's logotherapy or a Christian version of this, Christotherapy, among other schools. All these treatments are built on the conviction that latent keys exist, hidden in submerged data, blocked impulse or neglected or suppressed stories, by which fresh, healing versions of personal coherence can be discovered. Access to this unexamined or unconscious layer of experience lies in conversation. Thus all these schools are in some way confessional therapies. Their efficacy rests on the power of epiphany, the liberating insight into the raw phenomena of one's malaise and the perception of a new gestalt, a luminous, unbinding account, making sense of the events of one's life and mind. Psychoanalysis does much good; it saves lives. Yet the culture at the centre of which psychology commandingly stands remains profoundly anxiety ridden. Could it be that there is an "orphic" band of discomfiture that usual psychological practice does not reach and cannot report, even as conditions there send their perturbations rippling through the psyche? I think this is the case.

Prior to and feeding many of the distresses millions of North Americans take to therapists is a sense of deprival that could be called ontological loneliness. I wish here to examine the roots of this condition in the Western intellectual tradition and possible means of treating it. But I must ask you to allow me to take what might appear to be a somewhat circuitous route into this examination.

In his dialogue in *Symposium*, which we have already visited and will visit again, Plato presents a group of Athenian aristocrats who are enjoying a leisurely party throughout an afternoon. This gathering takes place in the garden of the poet Agathon, who just the previous day had won a major literary competition for his first work of dramatic tragedy. Almost all present are suffering vicious hangovers from Agathon's main victory celebration, which ended, after all, only a few hours before. As a result, they decide to drink lightly and for amusement have a diverting discussion on the nature of love to pass the time. The two figures who eventually play the most significant philosophical roles in the dialogue, Socrates and the rising soldier Alcibiades, arrive late, but nevertheless agree to join in the spirit of the afternoon with their own speeches about love, what it is and what it wants once they appear at the party.

The playwright Aristophanes speaks around the midpoint of the festivities, and for his account of love he reaches into a version of human prehistory. The very first humans were magnificent creatures, he tells his fellow symposiasts, proud, powerful and, as befits such extraordinary beings, they were perfectly spherical, as was their motion, because, as Plato tells us, they were like "their parents in the sky" (*Symposium* 190B), the planets and their orbits. Most of these human spheres were androgynous, with male and female elements; some, however, were solely female or solely male. They had two heads and faces and two sets of sexual organs and walked upright except when they wanted to move quickly or were excited. Then they would extend their eight appendages and roll like wheels, cartwheeling wildly. They are strange, homo-arachnid creatures, as poet and philosopher Warren

Heiti recently described them in a talk, though Plato assures us they were exceedingly beautiful.

In the greatness of their pride, these exalted creatures decided to make an assault on heaven and supplant the gods. The deities, of course, were disturbed by this act of hybris when they learned of the plot, but also troubled in their minds about how to handle the threat. They did not want to wipe out the human race—who then would worship them and offer sacrifice? Zeus proposed a compromise: he would cut each one in two the way people cut boiled eggs with long strands of hair. He then arranged for Apollo to turn the head of each half around so it faced the right way, then to gather the excess of flesh left from the divine surgery and tie it in a knot, which became the human navel. Each severed human was inconsolable about the loss of its other half afterwards and pursued this lost fragment obsessively. This, said Aristophanes, was the origin of love and desire and an explanation of the great power of these emotions. Zeus, by the way, threatened to split humans again should their earlier hybris return, so that they would be forced to hop on one leg and, so far, we have managed somehow to avoid this pan-species punishment.

The experience of selfhood, says Plato through the mouth of the playwright, is the experience of loss or incompleteness and a surge of cunning directed at filling the absence in one's situation. He has Socrates's mentor, Diotima, tell another story later in the dialogue that underscores the moral of Aristophanes's tale. This story also concerns the birth of love. On the day Aphrodite was born, the account goes, the gods held a great feast. Poros, the god of cunning or ingenuity, attended, of course, along with the other deities. Penia, whose name means "poverty," also came but was made to stay outside the gates and here she hatched a plan "to relieve her lack of resources" (203C)— she would lie with Poros once the god became drunk on nectar and conceive a child by him. And this is what happened. Like his mother, Love, or longing, is always homeless and poor, always wanting, but like his father, he is perpetually scheming ways to fill up his loss. He is their amalgam, abject, but perpetually ingenious—thus his unbroken appetite.

Love, which we treasure since it is "a great spirit" (202E), from the divine perspective, in Plato's Aristophanic story, is a symptom of an injury inflicted to maintain cosmological order. If love, one way the gods "mingle and converse with us" (203A), were to be utterly successful in pursuit of its ambitions, being again would be imperiled. The cosmological order, and an entry-level sanctity, requires us, Plato's account implies, to be in an incomplete state, our injury unhealed. This debility, supplemented by a never completely appeased desire for healing, assures an unstable but workable concord. This seems a strange form of talk, but it shows the crucial providential utility of unsatisfied, ever-in-motion yearning.

What is it that human beings most need? Aristophanes says the part of themselves that they've lost because of their self-love, their over-reaching search for metaphysical control. But there are other possible answers to this question, obviously. Socrates, a true Pythagorean, in the same dialogue, tracks the vector of human longing past sexuality through science to a union with the ground of being, which he calls the Good or the beauty beyond being. The world making sense and being beautiful means the Good exists and can be approached. And there are still other plausible accounts of human desire. It is possible, for example, that what we in the West feel most acutely disconnected from is the earth itself, that our lack is not amorous but ontological, chthonic. This separation, this loneliness, indeed can appear to be a central preoccupation in Western thought, where it is often disastrously read as an achievement, indeed a virtue of the most evolved among us.

We see such a split in Descartes's division of being into two unlike forms, *res cogitans*, thinking things, and *res extensa*, extended things. The two categories in the Cartesian scheme have little to do with one another, with the first being the vastly superior one, enjoying unlimited rights over what is not human or other than reason. We see a split at the ontological level as well in Kant's distinction between phenomenal and noumenal reality, the object as it seems to us and the object as it is in itself, with the latter being absolutely inaccessible to human intelligence because of the mind's basic structure.

In Marx, we note the deracination of the self from his or her place, one's traditional geography, as part of the inevitable and revolutionary iconoclasm of the bourgeoisie and an inescapable historical reality. In the actual separation of the person from her minutely known place, which has shaped her thinking and feeling, a liberating anarchism is at work. One is "freed" from the isomorphism of self and locale in the bourgeois revolution, indeed is brought to a higher state. Then there is Marx's claim that nature itself now no longer actually exists; nature "has been sucked into history" (476), he says, by the dialectical force operating through events, driven by the febrile, lawless energy of the bourgeoisie. This disappearance of the other-than-human, the view that everything is culture, is a tacit understanding in neoliberalism as well, equally a sign of an important historical advance, but there the relationship-destroying power separating humans and nature is some sort of economic evolutionism.

The most fundamental description of the human-earth severing in the Western tradition is to be found in the Book of Genesis in the story of the expulsion of Adam and Eve from the Garden of Eden. The first human pair are driven by an angel from paradise, a charmed place lit with trans-species intimacy—Adam names the animals, his partners in incomplete ways—and they are forced to wander in a cold, desacralized waste—"thorns and thistles it shall bring forth for you" (Genesis 3:18)—which is marked by distance, a non-cooperation, even antagonism, between humans and the other-than-humans. Here, in this dead locale, human integrity very quickly unravels and the first homicide occurs. Cataclysmic though these troubles are, in the light of later Christologies, they are seen to be the result of a "felix culpa," a happy fault, permitting the salvation of Christ, though the healing of Christ is never total and does not extend to uniting humans and the earth. It does not restore Paradise.

Then there is the essence of a particular form of rationalism much prized in the West, hyper-logical, detached, utilitarian, efficient, skeptical, managerial, technophilic, which insists on a rescue of persons from the apparent disorder of nature and even from the sprawl of spontaneous human life. Such separation in the tower of dispassionate logic is taken as a sign of full cultural maturity. This particular vision of reason stands at the core of the West's current self-identity.

Imperialism, which such rationalism makes virtually inevitable, a central Western undertaking from even before industrialization, itself breeds apartness: one feels obliged to control, exploit—master—where one is and to impose one's will on it. Such calculation and aggressiveness make intimacy with the earth impossible.

Such are the stories we tell ourselves in the West. These accounts—religious, ontological, economic, epistemological, historical—work in us at a deep level. They are embedded in the culture even beneath the effacing forces of secularism and philosophical indifference. They both describe and cause our alienation from the place where we live and from the earth itself.

It is part of our nature, a post-lapsarian, modern, fully rational nature, these stories claim, to be apart from the world, best of all to be above it all, a continentalist, globalist, politically transcendent. An imagined Los Angeles appears everywhere. This belief causes incalculable damage, both environmental and psychic. How to bridge the chasm and be re-united with our severed chthonic segment in the form of the earth itself, in particular our special part of it? Conventional therapies have little to say on this problem, and traditional philosophy has already made its position clear in its defence of human exceptionalism and apartness over the last four centuries.

Poetry, however, I think can offer some assistance with this problem. I don't mean particular poems—Wordsworth on the Romantic sublime or Whitman on the goodness of the land—but the nature of poetry itself, its raw elements, the basic forces at work in it. These offer the possibility of a treatment of this disabling loneliness. The structure of metaphor, for instance, the dramatic, transgressive fusion of thing with unlike thing, tenor with bizarre and revelatory vehicle, is an operation of unification within apparently random heterogeneity or ontological birfurcation. My head is a pumpkin; my love is a red, red rose; the wine-coloured sea; a will like water.

As well, at the heart of the lyric poem is a fixed gaze, an arrested look, which builds to a pitch of contemplative attention. What one is pinned to in this state is the thisness of one's subject, your beloved, this street, that afternoon light, that Douglas fir, and what grows from this fixity is affection, a quiet, if not a deference, before objects, and home-going relation. Noting thisness—*haecceity* in the language of thirteenth-century philosopher John Duns Scotus—re-sacralizes objects, re-invigorates them with a recalled beauty that thrills consciousness and throws us at the feet of this tree, this mountain.

There is a third therapeutic force available to a certain sort of poetry: when it follows the creative spirit, what an Ibn 'Arabian epistemologist might call the angel, the formative passion of your being, poetry is the wilderness of language. As with any other form of wilderness, it can give one—writer and reader alike—freedom and make one ready, in this new, light state of permeability, for friendships, amities with anything but especially perhaps the other-than-human. The linguistic wilderness instigates a hyperbolically adventuresome appetite, thus our capacity in a state of verbal delirium to savour what appears to be farthest from us. Like squirrels our minds can leap from the wilderness of language to the wilderness of trees and stones and deer and water and ferns.

Poetry is a flimsy device in the face of the great injury of the human separation from nature. Nevertheless, it must do its work. Poetry begins to seem less frail when it bleeds into the practice, the routine, of disciplined contemplative attention. Traditionally this interior stance is associated with monasticism in Taoism and Buddhism as well as in Christianity, but great stresses produce monumental alterations in psychology, as in geology, and climate change may make monks of us all.

This meditative stance should not be read as a shirking of any duty to address the material problems of environmental crisis. It is not avoidance of enviro-activism, a life of quiet thought supplanting action; it is not a political quietism. Environmental disasters grow out of a pancultural, long-lasting set of cognitive allegiances, are the final expression in fact of widely shared images of the self and notions of what a human being is and must do. The resolution of problems in the biota—if one can hope for such—is initially and perpetually noetic, dispositional and results in a trued or more true subjectivity and a return to the larger body of the world. We may learn to love and mourn for where we live and move.

The therapy offered by the attraction of metaphor, attention and linguistic ferality found in certain poetry is ascetical, ludic, transgressive, ecstatic and always sad (*penthic*): it shapes a heart that is vitally sad because it cannot fully have what it wants and this makes it want its desired object even more. It is not a therapy that urges confession, but causes one, as the Psalmist says, to "mourn like a dove." That is, it breeds contemplative longing, which enacts an ever incomplete, yet bracingly sufficient, healing.

✳ The pre-poem life of the poem, as well as the politics the lyric poem may inculcate, is an ascesis, part of which has been outlined here, a somatic and gestural discipline. I would like to take time to examine more carefully this behaviour seated in the apparatus of the poem and in contemplation that might return us to the world. Environmental problems, from climate change to pollution of salmon rivers, contrary to the popular view, are not at root political, technological, economic or even biological matters and to view them in such macroscopic terms is not just to be misled but it is to be defeated. Our biospheric difficulties start, as I have been arguing, with feelings and, where these have hardened, certitudes, some catastrophic. One of the most compelling emotions that now stir within us is the sadness I have been describing as chthonic nostalgia. This sorrow, I think, must be recognized and its momentum—its teleology, its watercourse—must be taken up as one's own, and in this you allow yourself to be delivered to this emotion's particular resolution, which will never be the extirpation of the sorrow you wish to treat. In the deprival is the form of the cure, as Plato knew. Instead of the end of this loneliness, the dynamism of this sorrow, this reaching, will become the chosen flow of your life.

Thomas Merton, in his novitiate, talks on Cistercian usages, the *habitus* of the Benedictine tradition, the somatic repetitions that make up Christian monastic practice and understanding, encouraged his young listeners to discover the nourishment of the polyform daemonic sites in the community where they had elected to spend their lives, Gethsemani Abbey. He argued that as one stands—physical and interior postures in communication with one another—so one knows.

"We can begin our study of monastic observances," he told his novices late in the 1950s, "by looking at the monastery itself...the place of peace where everything is ordered according to wisdom" (Merton 7). The site forms those in it who act "in the right way at the right times" in its various places, shapes them "by its silent instruction" (8). The proper placement of your body, how you enter or leave the refectory, the manner in which you engage in choir, says Merton, results in knowledge of the hidden God, a slow seeping in of illumination, a teaching shaped by the pneuma of a location. The place itself is complicit in the achievement of a true understanding of the world and a right relationship with it, but only a practice of appropriate action releases the pedagogical power in objects and places. Here is the genius within the Benedictine emphasis on stability: your cell, if you enact what it physically urges, will teach you everything.

Merton on Cistercian usages undermines the view that the path to truth is analytical objectivity; rather it suggests the route to clarity is fundamentally somatic, a discipline of walking, standing, gesturing in a manner shaped by attention, a manner mirroring the individuating meaning of the place, as you continue to savour it with your double set of senses, the physical senses and their inner equivalents, the ear of the heart, the mind's eye, the mouth by which we taste the beauty of objects in the world. You crouch in the thing by placing yourself beneath it, by gazing at, tarrying with, by your abjection and by occupying the sadness of one who has lost an essential part, a capping substance.

In the Quranic exegesis of certain philosophers, one's interior disposition and physical posture before even the letters of the text are essential elements in comprehension. Mullah Sadra, for instance, in his *Tafsir Surat al-waqia* (qtd. in Rustom, *Qur'anic Exegesis* 91) recommends a disposition of meagreness and humility, kneeling in fact or in mind, before the verses—"oceans saturated with gems of meaning"— as a means of grasping the great gravity of the Word. Imagine yourself as the primary addressee of all aspects of a passage, every prohibition, affirmation, promise, threat, lamentation, he counsels (Rustom, *Triumph of Mercy* 16). Something similar could be proposed as central features of a granular ontological *lectio divina*, which would be a politics and a course of correction of subjectivity, an assembling of a life. You gain access to the meanings and beauties of the not-human through decorum, courtesy, emptiness, availability, affection and attention. Your hunger may be a source of these stances. These comprehensions, the meanings, the beauties, come to you as you enact a physical, human translation of the charged site in how you walk, how you look, how you touch or do not touch what is before you. You do not dig these understandings out of objects. Your behaviour may or may not soften the occultation of things, opening their hiddenness.

As you stand and move in your place, you indeed are a sentience that has grown from it, corvids, fawns, cats, raccoons, oaks being others. Hear with your whole body. Walk as the crow walks, sit as the mountain sits. Let your body be a text carrying the meaning of the object into its movements, its voice; things will invite, then you must mime, give human shape to, the daemon, the *haecceity*, the domesticity (what-it-is-when-it's-at-home) of the place. The appropriate behaviours are a deep, sagacious taking in, a reading, of the Garry oak meadow and its April show of camas bloom on the west side of this mountain at the foot of which I sit, and read, garden, think, write and move.

TWO READINGS ON SNOW, TWO READINGS ON SORROW

Reading Gillian Rose

Stegner House, Eastend, SK, January 2018

BRITISH PHILOSOPHER GILLIAN ROSE remarks at the beginning of *Mourning Becomes the Law* on the strangeness of living "in a time when philosophy has found so many ways to damage if not to destroy itself" (1). This recent tendency has not ended thinking, of course, even though most of philosophy's traditional pursuits—"eternity, reason, truth, representation, justice, freedom, beauty and the Good"—have been discredited and jettisoned. Simultaneous with the rejection of what has struck some as the elements of metaphysics has been a search for new forms of ethics. "Yet no one seems to have considered," Rose observes, "what philosophical resources remain for an ethics when so much of the live tradition is disqualified and deadened" (1). How will thinking spend its time under such sparsely provisioned conditions, how will it continue in this state of snow? What behaviours will fill this void? Rose suspects they will arise from a pursuit of a novel innocence.

She herself holds to a position of exchange and transformation, which she discovers in Hegel, the philosopher she considered, in Andrew Shanks's view, "the *most decisively anti-totalitarian*" (Shanks 18). She locates the stance of neo-Hegelian conversational conviviality in what she calls "the broken middle," a place Hegel himself described as "the Bacchanalian revel" of Spirit "in which no member is not drunk" (*Phenomenology of Spirit* 27). This Hegelian inebriation arises from the fluid sundering and erecting of explanatory shapes, as philosophy "begets and traverses its own moments" (27). These undulations and twitchings of questing understanding are also experienced, even if briefly, as repose by those alive within them, since comprehension's various moments found worlds "where self-knowledge is just as immediately existence" (28). But, as *Phenomenology* amply demonstrates, truth appearing as "its own self-movement" (28) is a more accurate way of depicting the terminus of thought.

As a result, all is not delightful conversation and protean self-recognition in the procession of spirit: the middle, where transformations occur, is broken, disturbed, un-innocent, a "speculative Good Friday" (Hegel, *Faith and Knowledge* 191), where the one occupying it, the dialectical philosopher, is open to attack on all sides, even while she senses repeatedly the ground falling away beneath her. Yet, as both Damascius (71) and Plato dramatically claim (*Phaedo* 69c–d), philosophical wisdom begins with Dionysius torn asunder. Rose's position is an epistemology as well as an ascesis, a politics, of conversation. Her Hegelianism is a philosophical adventure like Plato's, a philosophical gamble supported by dialectical connoisseurship: the route to a moral, completed life is through exchange, the comedy—*comoedia*, "revel, carousal"—of the movement of the Hegelian absolute (*Mourning* 64).

✶ "Athens, the city of rational politics has been abandoned: she is said to have proven that enlightenment is domination" (*Mourning* 21) in post-modernity, Rose observes. Her erstwhile citizens, now disabused of reason's earlier claims, have placed themselves on the road to "the new Jerusalem, the imaginary community, where they seek to dedicate themselves to difference, to otherness, to love—to a new ethics" (21), little suspecting they carry within the uninterrogated rudiments of yet another city, the polis sponsored by capitalism, founded on private property and the view of the individual as autonomous, self-sovereign and relentlessly competitive. Critical rationality, now happily unmasked as an instrument of control, threatens the operation of autonomous self-determination and the private hoard in its material and spiritual forms, the virtue hoard, since it is capable of asking individuals shaped by capitalism's solipsistic proclivities to risk all in the formation of new solidarities or in the occupation of previously unsuspected, generative selves. Decommissioned reason offers no obstacle to the fulfillment of the sanctioned neo-liberal self.

Yet in *Mourning Becomes the Law*, and more extensively in *The Broken Middle*, Rose examines the lives of three modern exemplars who chose to become denizens of the "broken middle" and occupiers of the unanchored selves available there: Rahel Varnhagen (the eighteenth century), Rosa Luxemburg (early twentieth century) and Hannah Arendt (mid to late twentieth century). "In her own way, each of these women exposed the inequality and insufficiency of the universal political community of her day, without retreating to any phantasy of the local or exclusive community: each staked the risks of identity without any security of identity" (*Mourning* 39). Each furthermore was triply set apart, Rose notes, as Jewish, as women and as individuals who were critical observers. Varnhagen pursued a life in the spirit's "revel" through her Berlin salon, where Fichtean reforms of the state, Schleirmachean reform of religion, Humboldtian reform of education all were fostered. Luxemburg occupied a similar mediating, precarious position with her anti-Leninist leftism and suspicion of a socialism construable as the height of bourgeois morality.

Arendt occupied the unprotected middle of her time through her scholarship and political acuity. In particular Rose is struck by Arendt's reaching insights into the roots of totalitarianism in the simultaneity of "the declaration of the Rights of Man with the demand for national sovereignty" in the French Revolution (*Broken Middle* 219; Arendt 230). "The practical outcome of this contradiction [the same nation was at once declared to be subject to laws...*and* sovereign, that is, bound by no universal law—*Rose's interpolation*]," Arendt continues, "was that from then on human rights were protected and enforced only as national rights and that the very institution of the state, whose supreme task was to protect and guarantee man his rights as a man, as citizen and as national, lost its legal, rational appearance...lost its original connotation of freedom of the people and was being surrounded by a pseudomystical aura of lawless arbitrariness" (Arendt 280–81; Rose, *Broken Middle* 219–20). Under these circumstances, Rose observes, politics, in the form of lightly intentioned, reasoned discourse, becomes impossible, since the self and its unlimited becoming is rooted in the state and the state is rooted in nothing, thus answerable to nothing, "and it is the changing configurations of that impossibility from the French Revolution to National Socialism and Stalinism that constitute the 'origins' of 'totalitarianism'" (*Broken Middle* 220).

In the face of all this, each woman adopted an "unsettled and unsettling approach" (155) "which is not a 'position' because it does not *posit* anything"; each remained within "the agon of authorship," where each in their own way cultivated an "aporetic universalism, restless affirmation and undermining of political form and political action, which never loses sight of the continuing mutual corruption of the state and civil society..." (155).

✳ What is the innocence, standing on the newly cleared ground of philosophy, Rose finds untrustworthy? It is that fuelled by *"exultant revulsion"* (*Mourning* 54) in which the observer's "self-defences remain untouched," an emotional state quickened by a film like *Schindler's List*, in Rose's view, or by, perhaps, the audio program played on buses stopping at Krakow hotels to pick up sojourners to Auschwitz nearby, in which the word "fascism" is not uttered. The "identity of the voyeur" is left intact; self-examination is not encouraged, nor is theorizing about the roots of totalitarianism in bourgeois revolution. The banality of evil, as well as its ubiquity, is not considered. National Socialism, as it is presented in the buses' electronic guide, is a grotesque, special case, an unhuman "monstrosity." The auditor's innocence is preserved—is served—as the apparently truly anomalous behaviour of the unfathomably evil others is more deeply underscored.

Even more troubling than this faux innocence, for Rose, is the gradual elimination of civil society in the era of the apotheosis of the individual. "Libertarian ideology masks the concentration of legitimate violence in the centralized state, which, formerly was effectively delegated and disbursed across the quasi-independent institutions of the middle. And it masks *the unleashing of the non-legitimate violence of individualized civil society*, which is provoked by the systematic inequality arising from that concentration. Fascist movements seek *the monopoly of non-legitimate violence*: that is why they require the rule of law, which they also undermine" (*Mourning* 60).

The notion of universal law within nations means that such movements of the far right or left will have no rivals within the state in their exercise of non-legitimate violence. At the same time, incipient fascisms "seek to overturn the age-old impulse and wisdom of politics: that to guarantee my self-preservation and protection of my initially usurped property, I must grant the same guarantees to the persons and property of others" (60). With this correction to self-striving set aside, the conversation subsidizing it silenced by irony, suspicion, cancelling or indifference, Rose finds it possible to imagine "that states which combine social libertarianism with political authoritarianism, whether they have traditional class parties or not, could become susceptible to fascist movements." The utterly and constitutionally innocent are not required to make such concessions or guarantees of preservation to their neighbours, their status meaning they themselves have no need of them, and fascism's liberties are thus protected by their mistaken self-identification. Under such conditions, genuine and possibly corrective mourning cannot effectively work, but "remains melancholia" (64).

✳ Rose delineates two sorts of mourning at the heart of contemporary political life, one of which proves to be bogus—"aberrated mourning," unimplicating transfixion by unparalleled horror and a subsequent exit from history, a stance she associates with Heidegger in his romanticism and techno-skepticism and "inaugurated mourning," a changing state that goes on; in this grieving, the mourner is always troubled, continually hopes, imagining successive configurations in which the collective may somehow live. The latter sort of mourning, a redemptive *penthos*, a reconciling just act, Rose identifies as true and truing sorrow, associating it with Hegel's "comedy of absolute spirit" (71), intelligence's tumbling through history. Rose's "spirit" is not pneuma set against matter, not an extra-historical teleological force, not a totalizing impulse devouring otherness, not a blind power obeying dialectical law, not part of a pre-determined, "watertight" (72) system, but rather that which urges various political, cultural, artistic, personal, interior shapes that emerge as people persist in more or less continuous exchange, listening, hypothesizing, reformulating, reflecting. For sincere, philosophic interlocutors, this is a way of thinking of and participating in Platonic eros, an unfailing, ever frustrated, epiphanic reach. Or, as Rose suggests in the case of Rahel Varnhagen, it is "an extraordinarily modern Nicomachean ethic of friendship, *philia*...vicarious and precarious..." (*Broken Middle* 198).

✴ The comedy of the *Phenomenology of Spirit*, an interior slapstick of unsuccessful attempt after unsuccessful attempt at self-definition, is "the play of personae—the story of how natural consciousness acquired 'personality'—legal, aesthetic, moral—a story itself fitfully comprehended by philosophical consciousness which then proceeds unevenly through the stumbling blocks of personified aporia as each configured concept is mismatched to its object and corrected by a newly configured concept mismatched to its object, again—and then again" (Rose, *Broken Middle* 10). The talk, left to itself, is incapable of ending short of the restored community, which itself—Hegel was wrong on this—can be only asymptotically approached.

The *Phenomenology* is triunely ludic, this spectacle of thinking, grasping as fluid failure—first as the "drama of misrecognition" in which "aims and outcomes constantly mismatch" (Rose, *Mourning* 72). Then there is the performance of reason, "full of surprises, of unanticipated happenings, so that comprehension is always provisional and preliminary" (72). Thirdly there is the darkly comedic absence of Greek ethical life and in its place stands deception at the heart of legal status "where those with subjective rights and subjective ends deceive themselves and others that they act for the universal when they care only for their own interests" (72–73). All these effects on display in the *Phenomenology* comprise "the meaning of *Bildung*, of formation or education which is intrinsic to the phenomenological process" (72). It is the theatre of thought in modernity, unled, questing, trying a range of depictions, both external and internal, perpetually faced with the task of moving beyond the misapprehension of what is other as oppositional. "For the separation out of otherness as such," Rose says, "is derived from the failure of mutual recognition on the part of the two self-consciousnesses who encounter each other and refuse to recognize the other" (74) as an extended or sharp-sightedly remodelled instance of itself, the mind beautiful and terrible in its fallibility. "This applies," she adds gracefully, "to oneself as other" (74) as much as to an apparently opposing self-consciousness.

✳ Rose, by means of a trued reading of Hegel's dialectic, sought a resuscitated politics, a form of personal society, a "third city" (*Mourning* 11) made discernable through "re-invigorated, open-hearted reason," reason, I would suggest, in the service of philosophical eros, rather than apodictic proof. This polis, she wagered, would be found "buried alive beneath the unequivocal opposition of degraded power and exalted ethics, Athens and Jerusalem" (11), an opposition born of the conviction that the horrors of the twentieth century, the Holocaust pre-eminently, extinguish "Western metaphysical reason" (11). Such a community would pursue a *penthic*, rueful, compuncted politics meant "to work through the mourning required by the disasters of modernity, to acknowledge them as body by returning the spirit of misrecognition to its trinity of full mutual recognition, instead of lamenting these disasters as the universal 'spirit' of metaphysics, of the logocentric West" (76). This politics of a truly erotic reason would have as its model and goad "the comedy of absolute spirit as inaugurated mourning" (76). Such thought, a shape-changing exegesis of one's interlocutor's ostensibly fixed positions, could take place only in the broken middle; indeed, its operation would be the sole means by which this bacchanalian place, this dangerous, procreative place, could be constructed.

✳ The model of this conversational form of thought, style of speech, quality of attention, may be Hegel's ever-transmogrifying spirit, but Rose also urges that it allow itself to be influenced as well by a new "religious literacy" (*Mourning* 82), in particular an acquaintance with the cognitive style associated with midrash. This is understanding as exegesis in multiple forms, as opposed to "the domination of non-negotiable reason," a convivial, questing form of rationality rather than a declarative, totalizing one. Indeed, this hermeneutical methodology, "restitutive criticism" (98), in the phrase of Geoffrey Hartman, imagines "the political community exclusively in terms of ethical discursivity or conversation" (Rose, *Mourning* 95–96). From this exchange, an exegesis of past and present and a fluid, shape-changing construal of one's interlocutor's apparently fixed, self-rooting, self-distinguishing position is worked out to a point of mutual, non-reductive attention. Mourning before the unutterable joins itself to conversation, in which one deepens as spirit and political actor. "This work of mourning is the spiritual-political kingdom—the difficulty sustained, the transcendence of *actual* justice. Though tyrants rule the city, we understand that we, too, must constantly negotiate the *actuality* of being tyrannical…[A]s I follow the urgent and haunting voice of our dead from Auschwitz, to know and yet not to know, to be known, to mourn, I incorporate that actual justice in activity beyond activity" (122–23).

✳ This essay, I see now that I, returned from retreat, revise it at home, is about sorrow over loss of certain crucial cultural and epistemological devices—habits, liberations—that may have been extremely useful to us as we undergo the dissolutions of climate change, inching fascisms, pandemics and the struggle stadiums of social media. As Rosa Luxemburg reminds us, "every form of arbitrariness tends to deprave society" (Luxemburg 392; Rose, *Broken Middle* 210). The behaviour described here, born in cataclysm, presents a vivifying discipline. Rose's inaugurated mourning is available yet; the ingenuity dread can provide puts it in our hands.

Reading Damascius

Eastend, Saskatchewan, January, the air minus 40 degrees outside the frosted windows

Reading the essential Damascius, fifth and final *diadochus* or head of Plato's Academy, is a contemplation of mostly non-existent pages. Reading a version of Damascius is attending to the gossip of enemies and uncomprehending exegetes, copyists and students. But the cognitive ambition—the cognitive hope—of *lectio divina*, of contemplative understanding itself, matches poetry's: the tesseraic assemblage is perfectly sufficient. The fragment, even if it is disfigured by passing through multiple interpreters, gleams and touches by its sensuous, epiphanic light much else—indeed, a form of everything else—and this oblique illumination is understanding superior to a systematic causal blueprint.

The modern version of that compendium of Damascius's thought, *The Philosophical History*, gathered from two sources compiled five and six hundred years after his death, is the work of scholar and editor Polymnia Athanassiadi and appeared in 1999. Her interest in Damascius initially was less scholarly than it was personal, philosophically archeological, marked by a desire to see her home place with greater clarity.

> My interest in Damascius and in his Philosophical History was kindled by my desire to find out about the pagan communities of late antiquity and their spiritual posterity. The movement of this search has been largely retrogressive from roots which lie in the present, for the Greater Eastern Mediterranean, which has been my home for many generations, is rich in what a scholar might call 'pagan survivals,' but a local would perceive as the natural way of going about life and after-life. (Athanassiadi 9)

While finding her Damascian reading dissatisfying at first, his work so peculiar "with its little lives, its allusiveness, brusque halts and sheer sense of loss" (9), she nevertheless chose to offer a seminar on *The Philosophical History*, which, though it first met in a regular classroom, soon shifted to her home study and eventually settled in an unoccupied ancient Athenian house on the north slope of the Areopagus, a building that she and her students could reach only by climbing through a hole in a fence. They imagined that the place where they squatted and studied had been the actual residence of Damascius himself.

Athanassiadi's entry into Damascius was appropriately a by-blow, off beat, yet erotic, in the sense of being directed by a personal desideratum—the wish to see the Hellenistic residue of her area with more intimate philosophical precision. She had come to believe that the oddly arranged shard-like nature of the thought of Damascius as it had come down to the present matched exactly "the fragmentary and cryptic character which underlies Mediterranean religion in all its secular and mystical forms" (9–10). This formal echo underscores the philosophical importance Damascius assigned to "patria," the theurgical power of popular culture in certain cities and sites (59). But Damascius also is relevant to all of us now in the failing industrialized world: reading his surviving fragments today doesn't produce arcane learning but necessary contemporary advice since he lived with the certainty that the traditional thought of his day, a complete, tested, life-sustaining world-system, "stood on a razor's edge" (par. 150), as many suppose thinking now does. Damascius's tragedy seems a close version of ours.

✲ The 480s were an especially tumultuous time in intellectual circles in the Mediterranean world. Much of the battle between the Chalcedonian orthodox—Christ has two wills, human and divine—and the monophysites—he has but one—following the Council of Chalcedon in 451, initially seemed to leave Plato's Academy, and other places of Platonic learning, relatively untouched, even though Chalcedonians, monophysites and Neoplatonists often studied in the same Greek and Egyptian schools. Persecution struck the Hellenists first in 488 with the anti-pagan suppression in Alexandria engineered by the imperial envoy Nicodemus, supported by the local church hierarchy. It's likely that this terror was at least partly the result of the victory of the orthodox, formed by an unhappy reading of *alter Christus* ecclesiology, over what they saw as heresy in monophysite Christology. The subsequent ascendency of Chalcedonian supporters into ecclesiastical and civil positions of full power created a potentially dangerous climate: the conviction that the princes of the church spoke and acted as a version of the second person of the Trinity would bear a number of great temptations.

Damascius, born to a wealthy Syrian family, had come to Alexandria in the early 480s to study rhetoric at Horapollo's school, then an important centre of philosophical thought and rhetorical instruction and a gathering place for prominent philosophers like Sarapio and Isidore, Damascius's eventual teacher. Damascius witnessed Nicodemus's year-long repression, which was fiercely encouraged by the church in the form of the archbishop Peter Mongus—Damascius called him "a fixer"—and a nearby monastic community. The repression included the torture of Horapollo and the looting of local shrines, ending in the scandalous public display of their sacred objects and the burning of them. Damascius, who had hidden Isidore during this upheaval, moved by the stoicism of philosophers during this persecution, became converted to the study of philosophy and to an activism in its practice. "Men tend to bestow the name of virtue on a life of inactivity, but I do not agree with this view," Damascius observes in *The Philosophical History*.

> For the virtue which engages in the midst of public life through political activity and discourse fortifies the soul and strengthens through exercise what is healthy and perfect, while the impure and false element that lurks in human lives is fully exposed and more easily set on the road to improvement. And indeed politics offers great possibilities for doing what is good and useful; also for courage and firmness. (par. 124)

Damascius, following a pilgrimage to various daemonic sites—Ephesus, Samos—reached Athens in 490, where he found Plato's famous school much in decline. In 515, he became the fifth *diadochus* of the Athenian Academy, and, in an effort to revive the institution, he invited leading philosophers from Africa and elsewhere and built a large educational complex on the Areopagus's north slope. An overall preoccupation of his tenure was to emphasize the influence of the theurgist Iamblichus over the systematizer Proclus. When Byzantine Christian Justinian ascended to the imperial throne, he made clear his antagonism to both monophysites and Neoplatonists. In 529, in an edict aimed specifically at Damascius's establishment, Justinian ordered the end of all philosophical instruction in the city, thus ending the long history of the Academy.

In response to this suppression, Damascius decided to relocate the school of Plato to Persia, and he, at that time in his late sixties, along with six other philosophers, began a trek, across enormous deserts, to Ctesiphon on the Tigris and the court of the young tyrant Khusrau. They had thought they travelled to an enlightened and malleable ruler, but they were mistaken; Khusrau, they soon discovered, though he might have read many of Plato's dialogues, nevertheless was impetuous and cruel. Within a few years, the philosophers partially retraced their steps, settling in Harran, one of the few places in the Roman Empire where any sort of intellectual freedom was possible.

* *The Philosophical History* is indeed a strange book on a first encounter and not only because we have it in mildly questionable fragments. Its preoccupations are *patria*, forms of popular culture associated with particular sites, and *paradoxa*, miracles of saints and other fantastic tales. *The Philosophical History* thus is full of hearsay, bits of biographical recollections, travel notes, anecdotes, along with vicious denunciations, scattered seemingly at random. How can this be philosophy or the history of philosophy?

✶ Isidore, Damascius's teacher at Horapollo's school, we are told, was devoted to "spiritual travelling"—"if he ever heard of some extraordinary or sacred phenomenon, whether secret or manifest, he wanted to witness it for himself" (par. 21).

Isidore was also a superb hermeneut, reading with great skill the deepest layers of a text, yet he was at the same time suspicious of erudition—"I have indeed chanced upon some who are outwardly splendid philosophers in their rich memory of a multitude of theories… in the constant power of their extraordinary perceptiveness" (par. 14), but "within they are poor in matters of the soul and destitute of true knowledge." Part of true knowledge was what came from "holy prayer," chiefly dispositions, their relations and their uprootings; "when the soul is in holy prayer, facing the mighty ocean of the divine, at first, disengaged from the body, it concentrates on itself; then it abandons its own habits, withdrawing from logical into intuitive thinking; finally, at a third stage, it is possessed by the divine and drifts into an extraordinary serenity befitting gods rather than men" (par. 22A). Damascius looks for a supple spirit, as does Hegel. He found Christians "incurably polluted" (par. 20); nothing could compel him to tolerate their company. The tendencies—the stiffness of an official propriety, for one—arising from an attachment to an architectonic theology, which supplants mystical experience, may in part account for his antipathy to post-Chalcedonic Christianity.

✶ An important element in Damascius's methodology—one that drove Photius, an early transcriber, to distraction—is digression: a shifting, episodic structure proofs a reader against the consolations of system, chiefly certainty, the corrupting rest of philosophical orthodoxy and its inertia and the flashing delight in a consequent conversational combativeness. Damascius not only practices the discipline of digression, but pointedly announces the approach of his lateral leaps throughout *The Philosophical History* (par. 5A, 78F, 103A, 116A). One is shaken anticipatorily and actually from the following of an argument by such breaks and is placed in a position to be startled by yet another luminous shard. Understanding is *catanyxic*, breath-taking as a result; the reader is rescued from an eristic, victor-take-all position in philosophic conversation. The daemon in philosophy is unhindered. Further, as Athanassiadi points out, the *patria* and *paradoxa*, popping up in the digressive flow, clear stopping points in Damascius's method of digression, lend his book "an air of orality" (Athanassiadi 60), putting it close in spirit to the Socratic, wisdom-preserving suspicion of writing. System, abetted by writing, may be the main corruption to which a non-oral philosophy is susceptible since it somehow subsidizes a psychagogic amnesia.

✶ "Of all of his [Isidore's] contemporaries," Damascius notes, "philosophers and laymen alike, he was in the highest degree taciturn and secretive, yet he poured out his soul in the advancement of virtue and the reduction of vice" (par. 30C). Damascius's preoccupation with Isidore in the first third of *The Philosophical History* led nineteenth-century scholar Rudolf Asmus to believe the whole work was a biography of his teacher, but this is no more the case than that the Platonic dialogues are a biography of Socrates, even if they clearly show Plato struck by Socratic beauty and Socratic depths, regarding both as philosophically significant. Isidore presents a style and phenomenology that stand as a pedagogy: take these as constituting a final cause, a mind within a mind, and you advance in philosophy in the sense that the philosophical problems become clearer as the subjective posture under consideration, Isidore's practice of *disciplina arcani*, for instance, lifts and turns the observer.

"Strange though it may sound, with all his noble and austere dignity, he appeared charming to those who met him. His main preoccupation was with the common interest of his students; at times however he would temper his gravity with playfulness, teasing those who made mistakes so cleverly that the joke masked the criticism" (Damascius par. 30D). All dialectic in Plato builds from the unexamined eros of the interlocutor; all philosophy proceeds within the degree of friendship the dispositions of the interlocutors permit.

"His criticism seemed to be wholly justified, at least on an unbiased judgement; but for those who judged by common or ordinary criteria he often appeared to go too far" (par. 32B). Even if apparently extreme, his corrections, jocular or not, sought to shape the soul, in particular by releasing three crucial forces in it or creating conditions where these powers might step forth.

> All agree that there are three primary and essential principles for an inquiry which contemplates reality: love, industry and sagacity. Love is the first and greatest principle, the most wondrous tracker after all that is beautiful and good. [Then one needs] sharp and sagacious natural powers, capable of covering much ground in a short time, truly adept at following up and recognizing which of the quarry's tracks are genuine and which are false for the purposes of the chase. The third requirement is relentless industry…" (par. 33A)

which in its tirelessness and truth-scenting resembles Hegel's spirit. What drives the moments of interior, philosophic alteration is not dialectical force, however, but discernment, and this power builds through "good fortune," "the possession of the divine," since "sagacity and acuity are not the same thing as swift imagination or conjectural talent" (Damascius par. 33C).

∗ *Paradoxa*—extraordinary, daemonic occurrences—would seem to have no place in philosophy or even in accounts of the interior life, where a chaste reticence around "favours" is ideal. But some reports of such events in a life prove to be generous encouragements, Socrates's account of his friend's visit to the Pythia and the oracle's reply to his question of who possessed wisdom, for instance, or Teresa of Avila's *Interior Castle*, the visions of St. John, and Ibn 'Arabi's visitations of the Angel reported in his *Futuhat al-makkiyya*.

> I saw the baetyl [a meteoric stone] moving in the air, now hiding itself in the clothes of its guardian, now held in his hands. The name of the guardian of the baetyl was Eusebius, who said that at some point he had suddenly had a strange urge to wander away from the town of Emesa in the middle of the night almost as far as the mountain on which is built an ancient temple [destroyed on Constantine's orders] of Athena; he came with great haste to the foot of the mountain and sat down to rest as one does after a long journey. He then saw suddenly a ball of fire leaping down from above and a huge lion standing beside it, which instantly vanished. (Damascius par. 138)

∗ "He [Isidore] was rooted in the very purity of Platonic ideas," which he often construed in an unusual manner, perhaps, though Damascius does not go into this at length, by following his beloved Iamblichus, who held that not only the gods, but also daemons and disembodied souls were unreachable by the passions and therefore changeless (Iamblichus I.10.36), making the operation of the agent intellect, the visitation of the angel, an even more transfixing moment, for this power, that form of thought, was especially nimble and unfailingly acute. Isidore was particularly drawn to Iamblichus's prizing tutored intuition over discursive thought.

Isidore deplored "the present situation," that is the vicious anti-Hellenist suppression, together with the ascendency of organized Christianity, though he "did not wish to worship statues of the gods, but he was fast moving toward the gods themselves, who are hidden within, not in sanctuaries but in the very mystery—whatever this may be—of the completely unknowable" (Damascius par. 36A).

> And how did he move toward them, when they are of such a nature? By the power of extraordinary love, itself no less mysterious. What is love if not unknowable? Those who have experienced it know what I am saying, for it is impossible to describe it, nor is it any easier to comprehend it with the mind. Truth itself is in danger of being extinguished. (par. 36A & B)

Isidore was not a good lecturer, nor an extensive scholar, nor a particularly skillful logician; rather, as a pedagogue he sought to teach his students "by placing eyes in their soul" (par. 38B).

✳ Yet another sort of philosophical device Damascius employs is late antiquity's version of medieval monastic exempla. These prosopographical sketches of incidents, patterns of lives—moments of decision, spans of moral amnesia—may shock, provoking an abrupt turn in a life, may introduce a particular telos to striving, may shape discernment. Their nature as formational tools is determined by the state of the reader more than by the intent of the speaker or author, which should be determined by the student's readiness, if known. Phenomenologies must be matched with phenomenologies as closely as possible. And these exempla also present a social history of a time in which significant cultural change is taking place—thus the tones of loss, remorse and disorientation in this section of Damascius's book.

> Epiphanius and Euprepius were both of Alexandrian descent and were experts in the mystical rites established among the Alexandrians, Euprepius presiding over the so-called Persian mysteries and Epiphanius over those related to Osiris—and not merely these but also the mysteries of the god celebrated at Aion. (Though I can disclose the identity of the god, I will not write it down on this occasion.) Anyway, Epiphanius presided over these mysteries. These men were not born into the traditional way of life, but they overlapped with and met those who had and, having benefited from their company, they became for their contemporaries the source of many blessings and, among other things, the powerfully voiced messengers of ancient tales. (par. 41)

*

> Hypatia: she was born, brought up and educated in Alexandria and, being endowed with a nobler nature than her father [the mathematician Theo], she was not content with the mathematical education that her father gave her, but occupied herself with some distinction in other branches of philosophy. And wrapping herself in a philosopher's cloak, she progressed through the town, publicly interpreting the works of Plato, Aristotle, or any other philosopher to those who wished to listen. (par. 43A)

This Hypatia was beloved by the whole city, "[e]ven if philosophy itself was dead." Cyril, bishop "of the opposing sect," that is, the organized church, observing the great gatherings of men and women milling around her, in a state of envy toward her, incited a crowd of "truly abominable" men to beat Hypatia to death (par. 43E).

Then there was Hierocles, who famously once declared that Socrates's words were like unusual dice: however they fell, they fell rightly. He did not mean to imply that Socrates had magical powers that controlled chance, but that with certain phronetic speakers one could begin anywhere and begin perfectly. Hierocles's "brave and noble nature" was demonstrated by the fate that befell him.

> Once in Byzantium, he gave offence to the ruling party [the Christians] and, being taken to court, he was savagely beaten up. As he flowed with blood he gathered some of it into the hollow of his hand and sprinkled it over the

judge exclaiming: "There Cyclops, drink the wine now that you have devoured human flesh" [*Odyssey* 9:347]. (Damascius par. 45B)

✳ How does Damascius accommodate this mood of loss found throughout his book—the loss of a world-system deeply informed by scholarship, contemplative practice and theurgy, the loss of this aquifer of life-meaning—and violation? He remembers and reports. "Divination by clouds, which was not known to the Ancients even by hearsay, was discovered by a woman called Anthusa at the time of the Roman emperor Leo" (par. 52). He admires wisdom's underground, unnoted life, its green innovation.

> This woman came from Aegae in Cilicia…Concerned about her husband, who had been invested with some military office and sent with others to the Sicilian war, she prayed to the rising sun for foresight into the future in her dreams. Then her father, appearing in a dream, exhorted her to pray to the setting sun as well, and as she prayed there formed out of the clear sky a cloud around the sun which eventually grew and took the shape of a man. Another cloud broke away from it and grew until it reached an equal size and took the shape of a lion who became angry and, opening its mouth wide, swallowed the man; the man-shaped cloud looked like a Goth…From that time until now Anthusa has continued to practice this method of divination by clouds. (par. 52)

One reads with care what steps forward, what commands attention in clouds, plants, water, ceremony; all is *lectio* and merits careful, lively interpretation.

Philosophy for Damascius, in the time of the exile of philosophy especially, is not a specialized activity; it lives in the village and is parsed at the communal well. Here it is not bookish but intuitive, nature-reading and theurgical. Anecdote, phenomenological report, stand in the place of the more metropolitan activity of synthesis. Thus Damascian philosophy, philosophy in the twilight of philosophy, is local, intimate and vital, buoyed by private and communal conviction, passed on by story. It has a strength that does not come from organization or institution. It does not forestall disaster, or even survive it, but, in Damascius's view, is a graceful way to meet it.

IN THE TIME OF EXTREME HEAT, IN THE TIME OF THE DISCOVERY OF UNMARKED GRAVES AT THE SITES OF RESIDENTIAL SCHOOLS

11

First Words

What is not mentioned in "Interiority and Climate Change," and other essays I have written on psychagogy and global warming, or in "Contemplative Practices, Contemplative Pedagogies," because I had not previously truly known it, is the deep terror experienced in the face of extreme weather—extreme heat, precipitation, drought—the engulfing power of these marks of the weather apocalypse. These eruptions spark dumbfounding; what comes stands as indifferent to life, certainly known, regular human life, the suburban life of the last sixty years; and what seemed solid before, the persistence of trees, birds, animals, does not now. These weather events are strange and savage like the old Hellenic gods.

✶ Another urgent matter, though perhaps not everyone's—Roman Catholics must identify what attitudes in Catholicism instigated the vicious, thanaphilic culture in residential schools that religious orders ran over a hundred-year period in North America. These dispositions, missiological, ecclesiastical, spiritual, inter-personal and the thought-worlds backing them up, must be purged: much will disappear if this exercise is performed with conviction. Parts of hegemonic whiteness will be disabled and a certain form of religion will become uncomfortable to practice. The roots of all politics, the roots of such deep cultural failures as these two, climate change and ecclesiastical complicity in colonialism, lie in epistemology.

I am with these thoughts in the summer of the discovery of unmarked graves in Kamloops and elsewhere and the west coast heat dome, weeks of doubled dumbfounding.

✳ A correction in settler sensibility that goes deep in, it is now even clearer, must be achieved. This work can be done in rarely visited corners of the tradition this sensibility claims. I am drawn to a particular teaching style, a form of address or dialectic, and a style of learning, in these political times, as I have proposed certain outsider forms of reading (*Going Home*) and of looking (*Living in the World As If It Were Home*) in the past, as medicine to treat the psychagogic poverty brought to North America by Europeans. This poverty that has been the basis of a monstrous culture, an instance of a tearing force that in SENĆOŦEN is called S̱ŦÁLEK̲EM. I will circle the pedagogical style of Plato's Diotima, the liturgical theurgy in pseudo-Dionysius's *The Ecclesiastical Hierarchy* and Maximus the Confessor's thoughts on contemplation and justice to approach the malaise in the belief that a cure lies somewhere roughly within the precincts of the disease.

Religion, or at least some sort of spirituality, must be part of the discussion of the human response to climate change, but religion of what sort? And if one of the dominant religions of the recent past has proved to be a moral and political failure, what options do we have? I see the Dionysian tradition as a way through this conundrum. By this tradition, we can be helped to join the community—beautiful, sustaining, tender—of things, Douglas fir, hummingbirds, arbutus (K̲EK̲EYIȽĆ). Named correctly, these presences can note us and take us in. But we must change before they can see us by means of the same gesture by which we truly see them. The weather effects of climate change seem religious—apocalyptic, biblical, the end of the world, we say of them—and the response to them must have some of the same flavour.

Reading about residential schools in the aftermath of the discovery of 215 unmarked graves on the grounds of the former Kamloops Indian Residential School and 751 similar graves on the Cowessess reserve in eastern Saskatchewan and more unmarked graves on the grounds of the former residential school on Penelakut (Kuper) Island, I ask myself: how could they have done this? "They" includes people like me, diasporic Europeans, who come from some of the same shaping forces, Cartesian, hyper rationalist, acedia-driven, that gave birth to a form of Christianity that weaponized the institutions. People wait for a papal apology that shows no sign of coming in the summer of 2021. But more than an apology is needed. What is required is a theodicy. I gather these thoughts in the tent "far off from the camp" (Exod. 33:7).

Diotima's Instructional Style

The first mention of Diotima, one of Socrates's earliest teachers, appears in *Symposium*, the dialogue in which we have spent much time, right after Socrates has questioned Agathon, the charming, award-winning author, a cresting celebrity, in whose honour the luminaries who people the dialogue have gathered. She is from Mantinea, we are told, "a woman who was wise in many things" (Plato, *Symposium* 201D). And she is powerful: she kept a plague at bay by telling Athenians what "sacrifices to make." She is a ceremonialist, then, a theurgist; she is the one who taught Socrates the nature of love, the single thing about which he claims to have knowledge. Socrates proposes to repeat her speech on this topic given to him when he was at the beginning of his philosophical life to those gathered at the party. He is interested in her teaching style as much as he is in her content— "I think it will be easiest for me to proceed the way Diotima did and tell you how she questioned me" (201E). Take this style I show you as a heuristic in all your inquiries, Socrates suggests to the lounging and drinking friends; note that this instruction is therapeutic; any ontological point made in it will come side-by-side with, or through, a moment of turning in her interlocutor. This performance of repetition will be Socrates's contribution to the afternoon's discussions on the agreed-upon topic of the nature of eros.

Even though his focus is Diotima's style, her teaching way, Socrates's exchange with her begins with the young version of himself giving an extended account of his own views on love. The young Socrates has behaved this way before, in *Parmenides*, when he blurted out a long treatment on cosmology before two masters, Parmenides and Zeno, jumping in before they had a chance to speak. Socrates's speech on love is full of the same puffy rhetoric, generalizations, as Agathon's was—Love is a great god and belongs to beautiful things and so on.

Diotima begins her instruction by shocking him from the torpor of received opinion, in this case a sweet piety about love: Love is neither beautiful nor good, she tells him. She severs him from sentimental common sense. Socrates mistakes Diotima's opening remarks for an eristic move on his teacher's part, and he resists her turning-the-soul-around probe with a quick ad absurdum reply—is Love ugly and bad, then? She quickly reminds him that their conversation is within the realm of spirit and he should not give his language such generous licence. She warns him not to demean the quest by taking their conversation to be a debate. The stakes here are higher than winning an argument: she tells him to watch his tongue. She also challenges him to be not so lazy in his thought; love's state is not an either/or matter; there is a place between beauty and ugliness as there is between wisdom and ignorance. It is a sort of spontaneous discernment or phronesis, "judging things correctly without being able to give a reason" (Plato, *Symposium* 202A). It's clear from the early moments of their encounter that the young Socrates doesn't possess this prelinguistic ability. He is not a connoisseur of good judgement; his reach is not deep or subtle enough to give him this power, his contemplative habits not formed.

Diotima then proceeds to lay bare Socrates's latent, unknown-to-him position, lifting it from potency. He has just declared Love to be a god, but he's also claimed the "god" Love "needs good and beautiful things" (202C), even as he admits that gods, by nature, are beautiful and happy. "Then how could he be a god if he has no share in good and beautiful things," she asks (202D). Socrates, in fact, holds the view, without realizing it, that Love cannot be divine. She combs the burrs from his instinctive position, and in so doing, she uncovers another band of being, the layer of reality that is constituted by the daemonic, the forces, "messengers," that travel between the gods and mortals, "conveying prayer and sacrifice from people to gods, while to men they bring back commands from the gods and gifts in return for sacrifices." Through intermediary powers like love "all divination passes, through them the art of priests in sacrifice and ritual, in enchantment, prophecy and sorcery" (203A), a mind before mind. The gods rarely mix with persons, but we can move in their air. By uncovering the sphere of the messengers, she makes it possible to see, value and live in response to a domain within the reach of divinity; you do so by taking up a practice of patient attending to its denizens. The ontology Socrates holds to be true and the philosophical imagination sustaining it are expanded by this revelation of an additional sphere of being, while his sense of self is decentred even as possibility grows. A previous self begins to break down, and in the collapse, Socrates is immediately upended and is delivered into the hands of one who can actually help him.

After he admits to this extension to being, the addition of the daemonic realm, the "many and various" spirits who are the instigators of insight and virtue, he oddly asks about the parentage of love. Everything has gotten richer for him but he experiences this plenitude as disorientation, loss, a state his teacher will now describe from multiple angles. Diotima replies she will tell him about love's ancestry, but warns him that the tale is long. She recounts the story of the meeting of Penia (poverty) and the drunken god Poros (way, resource), son of the goddess of cunning, both of whom attended a celebration organized by the gods to mark the birth of Aphrodite. Late in the festivities, Penia "schemed up a plan to relieve her lack of resources: she would get a child from Poros" (203C). Thus begins a capacious, visionary, formational tale meant to draw Socrates to a deeper understanding of the forces at play in him and a re-evaluation of certain seemingly undesirable states. The intent of the story is therapeutic rather than theological or ontological: Socrates now has an extraordinarily creative power, love's emptiness and diligence, an apparent non-power, in him described and thus this power has now become occupiable: soon he will see that it is the main channel of an inevitable and unrefusable erotic life. The daemon love blesses him with what it has received: poverty and unstinting ingenuity. Diotima, through her spiritual direction, brings Socrates to the point of this valuing of two aspects of essential absence.

✷ When the heat rose to unprecedented temperatures this summer on the coast, everything seemed to go quiet, locked in a kind of sick anticipation. The air was still, muted, birds disappeared and plants were scorched. People walked—few walked—from one pool of shade to another. While the heat dome lasted little more than a week, there was a sense that this extreme heat couldn't be borne much longer. Then the great fires began.

Ambiguum (Difficulty) 10

The seventh-century *Ambigua* of Maximus the Confessor are brief notes arising from Maximus's discussions of obstacles in the writings of St. Gregory the Theologian (Gregory of Nazianzen) and one exchange on difficulties found in the works of pseudo-Dionysius the Areopagite. Two gatherings of *Ambigua* exist in distinct collections, one where the interlocutor is John, Bishop of Cyzicus, and the other addressed to a certain Abbot Thomas, Maximus's later spiritual guide. The two works have circulated as *Ambigua ad Joannem*, composed likely in Maximus's African monastery, 628–630, and *Ambigua ad Thomam* from around 636. John Scotus Eriugena's translation of the *Ambigua* in the ninth century includes only the first collection.

   Salient interests pursued in the discussions between Bishop John and Maximus include the makeup of the human individual and the meaning of divine providence or mercy. In *Ambiguum* 10, Maximus argues against the view that the mind can reach God, become repeatedly aware, say, of numinous presence in pine needles, nuthatches, through reason alone without the assist of a period of ascetical wrestling and alignment. The incapacity of reason is twofold—following Gregory of Nazianzen on the tarnished state of post-lapsarian thinking, Maximus sees that "every human mind has gone astray and lost its natural motion" (Louth, *Ambiguum* 10, 1112A–B), meaning that the mind, in its disarray, acts as a cloud or veil, so that reality, sensible and metaphysical, is coated with incomprehensibility at a certain level as far as analysis is concerned because of the operation of analysis. But even in an uncorrupted state, reason alone would be incapable of complete understanding and effecting "participation in the Good" (1108C); it does not have the power to satisfy knowing's natural ambition for further and further clarity, leading to intimacy with the centre of this light. Aside from reason, there is contemplation and, standing beside this noetic engagement or, better, operating within it, steering it, is ascetical practice.

A particular behaviour creates access to truth, so that you may become "portions of God" (1108C) as an act of knowing. "For the contemplative cleaves to truths rationally and with knowledge, not with effort and struggle, and, apart from these, he refuses to see anything else because of the pleasure he has in them" (1109C). Reason, leagued with contemplative savouring shaped by an ascetically formed interior disposition, which includes a stance of patience in observation, for instance, an inclination to attentive play in the cognition of things, a permeability to beauty, gives you the world as it is, and this vision makes for delight.

✼ Capitalism and colonial hauteur place an additional thick film between human interiority and rocks, trees, shores, this patch of moss, deer antlers all but covered by dry grass, "God who is and appears through all things and in all things" (1116A), making respect and praise for these things difficult. Before this theophanic spectacle, people should "gather for themselves every capacity for wonder and reason for glorying" because of the pleasures of this wonder, one of the chief being becoming the recipient of the mercy of the land, another pleasure resembling Moses's "participation in glory" (1117C), a walking in the numinous sense of places and things. The great coral rose, in the hot corner of the yard, falls forward through the weight of bloom, contriving, in its bent state, to send out flowers from its back. Journey "hiddenly" to the intelligible world through the exercise of dilated reason amplified by imagination, eased by ascesis, discernment and apophasis as a form of deferential intelligence: this performance is the Mosaic rod, says Maximus, that with a single blow parts the waters of the constraining sea.

Maximus's anagogic hermeneutics, based on a concession concerning, and an affective probe into, various ontological layers, resembles a Khidr-like, Ibn 'Arabi-like *ta'wil*, or deep reading, of events, one's trains of thought, the past, texts and things. This lively poring over the imaginations of prophets, the withinness, godedness, of non-human life, speeding hummingbirds at play brushing against one another and cutting away, is an epistemology admitting the action of a previous, autonomous and providential mind, visits of the angel, Osip Mandelstam's whisper at a distance with its "astonishing independence" (3). This mind stretches the present deeply into the past and, through ascesis and *conversatio morum* (conversion), into the future and makes these domains part of the essence of one's self. This form of plunging reading further builds the self into others and such objects as trees, cliff faces, meandering deer, the sun and the stars. These interior behaviours, it is my hunch, are the remains of the old Neolithic chthonic rites and sensibilities, which apprehend luminous presence everywhere, see everywhere its creativity at play, an attitude displayed, for instance, by the astronomical and riparian designs carved on the back corbel stone in the tunnelled mound of Newgrange (Lewis-Williams and Pearce 228–30).

Ceremony and Thinking in Pseudo-Dionysius

The sixth-century book *The Ecclesiastical Hierarchy*, addressed to "Timothy the Fellow-Elder," contains an examination of "[o]ur hierarchy," which "consists of an inspired, divine and divinely worked understanding, activity and perfection" (Pseudo-Dionysius, *Ecclesiastical Hierarchy* 372A), that is, it holds an epistemology, as well as a program of self-cultivation and a vehicle of transformation. The author of *The Ecclesiastical Hierarchy* will attempt the triple revelation of these powers with "the aid of the transcendent and most sacred scriptures" and a study of "the sacred mystagogy," by which he means the capacity of various rites to illuminate and change participants. His instruction is non-missiological; indeed, it is urgently private: keep these potent psychagogic matters secret, he cautions his reader.

Through ceremony, pseudo-Dionysius says, we are made more available to "those beings who are superior to us," who are existent, yet "conceptual" (373A) and "out of this world" while being insistently present, muse-like, the angel understood as non-intentional cognitive events and the agent intellect these messengers transport. The ceremonies and the energies they foster help us with "that yearning for beauty which raises us upward (and which is raised upward)" (372B) to him, logos, who "pulls together all our differences."

Then, "[f]ormed by light" (372B), initiates in theurgy, "we shall be perfected and bring about perfection," and be "brought as far as we can be into the unity of divinization" (373B), that excellence, which is a matter of seeing the world as it is, a flowing multiplicity of communicating thisnesses, idiosyncratic angles of stems, animal gazes. By hierarchy, pseudo-Dionysius means "a state of understanding" that comes as close as possible to the divine (*Celestial Hierarchy* 164D), a knowing which is, among other things, a community with divinity in this ginger fern, this patch of salal, this thick-barked Garry oak I look at every morning. Such looking feeds a disposition that amounts to a freeing of truth so that a person is engaged with this everywhere-present indwelling spirit through "inspired participation" (Pseudo-Dionysius, *Ecclesiastical Hierarchy* 376A). In ecstatic speech, poetry and prophecy, as well as in careful looking, "something united" is made variegated and plural and the transcendent is brought "down to our level," divinity now being an acute specificity in things.

Echoing Diotima, pseudo-Dionysius says, "it is love of God which first of all moves us toward the divine" (392B), but it is not eros alone that has the ingenuity and power to complete this *peregrinatio*; mystagogy is also required since "there is risk for us when we handle what is above us" (392C). Ceremony provides seclusion, which is a defence against reductionism and overreaching and an aid to insight through the shaping of behaviour: *ah, so this is what I am or should be; this is how I must act.*

✴ Duende, the agent intellect, the Ibn 'Arabian angel, the acted-out interest in us on the part of other species, the possibility of meeting Khidr, make life worth living and ward off smotherings of the self through opinion and ideology.

✴ Engagement with a Diotima-like conversationalist, one who has travelled much of the erotic route and is inclined to speak about the journey, is as important as ceremony and its mouldings. But presumably this conversation will take place only with those who have been previously "snake bit" (Alcibiades's description of the experience) by an overpowering encounter with beauty; you, in your aching lack, ask "to be brought to the hierarch" to whom you promise "complete obedience to whatever is laid upon" you (Pseudo-Dionysius, *Ecclesiastical Hierarchy* 393B). The absence you feel—note how elation and emptiness elide in an experience of beauty—makes for this discerning availability to the guide, but, importantly, you do not pursue your case with just anyone: think of Teresa of Avila's breaks with several dismal directors. Your hunger needs a dependable, while general, phenomenological pattern arising from the experience of another to encourage and vivify your imagination—this pattern will embolden you and make you interiorly supple. You take the broad phenomenological idea and momentum from the guide, a "training" that is common, but also carefully fitted to the particular person and to different versions of that person. Think of Diotima's latitude with Socrates in the initial stages of her encounter with him and her severity after. Socrates eventually listens carefully to his teacher; growing from this attention he develops the phronetic inclination he previously lacked, and, as a result, he finds little coming from Agathon to be worth serious study. There may be a go-between in the dialectical enterprise, one who brings the one "fired by love of transcendent reality" (393B) to the one holding the title of hierarch; pseudo-Dionysius discusses this ascent in the context of a reading of the rite of baptism where the go-between is the sponsor. The one who brings the thirsting other is marked by "fright and uncertainly" when he thinks of "the heights" desire would have his charge scale.

The one so brought, the one aware of the absence he carries, wishes to recover his "lack of knowledge of the truly beautiful, the absence within himself of a God-possessed life" (396A), God being in significant part the multi-layered, merciful world. "[S]acred symbols are actually the perceptible tokens of the conceptual things" (397C), observes pseudo-Dionysius. The appropriate gesture in liturgy creates illumination. The body, under careful directions, liberates the mind. But eros does not always strike an unwavering, unerring route: one can "fall away from the life of the mind" (400A) and "remove themselves from this light." But the light "hastens to follow them."

✳ I cannot claim that Christianity extinguished the mystical, lost that beauty that is an encounter with beauty, and became a ready servant of the gobbling state in the time of imperial expansion, from the seventeenth to the nineteenth centuries, and even earlier in the suppression of original Mediterranean spirituality in the sixth century. The thought, the interior re-making, of many, John of the Cross, Marguerite Porete, et al., does not allow an unqualified version of this appraisal. But I clearly see that the esoteric was kept at the fringe of the official church, occasionally coming under suspicion, and that at times the boldly mystical was silenced. Porete was burned at the stake for her book *The Mirror of Simple Souls*, John of the Cross was imprisoned by his confreres because of the lyrically inventive zeal of his reforms, and Teilhard de Chardin was forbidden to publish on philosophical and theological topics by his superior general in 1947 because of the flowing verve of his metaphysics and his evolutionary notion of being. The end of the conversation between the church and Platonism, with Justinian's church-encouraged suppression of the Academy in Athens during the headship of Damascius, also marks a narrowing of contemplative thought in the church and the pursuit of fortressing doctrinal rectitude. Benefits associated with extended dialectical exchange, vigorous exegesis and attention to the play of idiosyncratic formational eros were harder to come by when this link was severed. The church, co-terminous with the nation, became over the years an institution bound up in the resurgence of the Roman Empire, an institution that could, in time, make deals with colonial governments to assist in the assimilation of Indigenous children so that a totality could be created by means of a cultural genocide.

Partly this monstrosity is the result of a way of life arising from certain absences—the absence of a certain form of pedagogical conversation, illustrated in this essay by Diotima's formation of Socrates, for instance, a way of understanding through theurgic engagement in ritual and a way of thinking shaped by imaginative hermeneutical boldness, ascesis and a creative theandric collaborative acting, all making up the antipode of a more acquisitive epistemology. Then there is the Eurocentrism and racism of even the most sympathetic of the early missionaries—Brebeuf, for example.

This is not to say that a more Dionysian church would have lacked doctrinal clarity in its conversational engagement with peoples, which likely has a Pauline root, but such a church would have stood a better chance of being authentically curious before new cosmologies held by different peoples. It would have been marked by an ecumenical appetite for learning from other cultures as the early church had learned from forms of Platonism. Such a church through history may have been more drawn to continued reciprocal conversation. An energetic creative complicity in thought could have led to a more welcoming, more widely savouring, ear. Instead, the church chose to base itself on an architectonic system in theology, resting on a certain form of reason rather than on mystical experience; in its evangelization, it chose totality over conversation, in liturgy rich display over theurgy.

The church, over various pontificates, became a body extremely unlikely to criticize the state, or to stand significantly apart from it. The institution may have appeared to change with the rise of left-wing Catholicism, with its base in liberation theology, and in the heightened ecumenism modelled by Thomas Merton and others in the years following Vatican II. But the aberrational status of this radical option, as far as doctrinal conservatism was concerned, was demonstrated repeatedly during the long left-suppressing rule of Pope John Paul II.

Nevertheless, I believe a resuscitation of the epistemological commitments of the Dionysian way are an important means of de-totalizing Christianity and making it a less toxic interlocutor—a listener almost entirely, like the young Socrates was eventually to Diotima—with resurgent Indigenous thought. Occupying the attention of the one being taught is the primary personal and political task. Socrates allows his teacher to take him and his views apart so that he can examine himself with greater honesty. The Dionysian portion of the contemplative tradition, enriched by Platonic and Neoplatonic examples, subverted by a magisterium intolerant of any syncretism, was minimized in a church committed to serving a dogmatic orthodoxy and intellectual hegemony. It is time to disinter this old ecstatic way.

Ambiguum 10, On Discernment

> Either the human community must offer a structure in which esoterism is an organic component; or else it must suffer all the consequences implied by a rejection of esoterism. (Corbin, *Creative Imagination* 15)

Maximus pursues here a contemplation of the prophet Elijah's vision at Horeb (1 Kings 19:9ff) in order to explore and implant, or reinforce, in his reader the power of discernment, an ability to detect the genuine. His use of the word "contemplation" signifies his *lectio divina* reading of messages, deposits of the messenger's mind, in stories and lives—allegory, but this at daemonic concentration. One is met by epistemological effervescence sequestered in these places, these anchorages of insight. "So Elijah is shown to be most wise after the fire, the earthquake and the wind that rent the mountains, which I take to be zeal, discernment, and an eager, assured faith" (Louth, *Ambiguum* 10, 1121C). For discernment utterly alters the ingrained habit of evil, "assaulting it through virtue like an earthquake breaking up what is held together" (1121C–D). Faith, the breaker of mountains, disables "the trains of thought and the demons of sophistry" (1121D) by which evil abounds. These actions gain you access to the last manifestation, utter silence, the numinous "voice" before which Elijah, the one prophet in Israel not killed by Ahab and Jezebel, stands "wondering at its glory and wounded by its beauty," a particular state of being, precarious, vivified, which he is driven to emulate "rather than just pursue" (1124A). He passes through matter in "the divine chariot of the virtues"—he's sped by the disposition, achieved by the upending of phronesis—and finds a passage to the intelligible realm, that world in or beside the world. The loss of the interior acuity fashioned by a contemplative training like that of Maximus, a mystical form of life, makes a fine reading of what appears unlikely, blocking the full, affect-laden instruction of texts, things, places, experience and the views of others, the route to expansive beauty, making moral beauty rare. This loss makes it difficult to take in the world as it wondrously is.

Ambiguum 5, On Theandrism

Maximus begins this *Difficulty* by quoting from pseudo-Dionysius's fourth letter, addressed to the monk Gaius, on the confluence of transcendence and "being truly a man" in the person of Christ (Pseudo-Dionysus, *Letter* 4, 1072A; Louth, *Ambiguum* 5, 1045D). Maximus was a Chalcedonian—the Logos possesses two wills—with the assist of Dionysius's Christology.

If other human wills, all human wills, are to participate in this confluential enterprise, human acts sharing the same streambed as the possible transcendent act, these wills need to be shaped by a rich ecstatic imagination, an imagination that is the result of ascesis, theurgy and a discerning extravagance, commensurate, or at least on speaking terms with, the transcendent will. "[T]he great Dionysius corrects the monk Gaius with these words, teaching that the God of all, as Incarnate, is not simply said to be man, but is himself truly man in the whole of his being" (1048A) and that this truly human nature somehow shares a psyche with divinity. If this is the case, ideology,

Plato's Great Beast, powerful as it is, has no complete control and gorgeous worlds can be worked out or rise from human imagination. This visionary effort, acting complicitly with the one not "bare God 'but one who is in different ways truly man and the lover of man,'" the logos as a second yet primal mind (1048C) is the natural end of all consciousnesses. "For out of his infinite longing for human kind he has himself become by nature that for which he longed" (1048C).

The Logos's humanity, while being truly human, can be approached, however, only apophatically, in craning wordlessness, "for he himself remains himself completely incomprehensible, and shows his own incarnation, which has been granted a generation beyond being, to be more incomprehensible than any mystery" (1049A), a hiddenness in revelation, "spoken" yet ineffable, an abundance of "transcendent being" locked into the particular, the oceanspray reddening in rising heat on the west slope of the dry mountain. This unsayability is the beyondness in all selves, the ultimate opacity of self, the clouded presence of the daemon. Maximus's instincts, nevertheless, are consistently Chalcedonic—"It follows then that it is necessary reverently to confess the natures of Christ, of which he is the *hypostasis*, and his natural energies, of which he is the true union in respect of both natures, since he acts by himself congruently, monadically, even as a single form, and in everything displays without separation the energy of his own flesh together with the divine power" (1052C). Such is the collective human vocation, this hybridity in fluent act. This mixing of inclinations is both modelled and powered by Logos, who "passes through what we suffer by nature, and by the authority of his intention he shows that what we can naturally move by our intention is moved by himself" (1053C), this consubstantiality in act a heightened form of human reach, the mind, for instance, in the act of composing Riel's *Massinahican*. This fused creative momentum is a demonstration of what pseudo-Dionysius calls "theandric energy" (*Letter* 4); this capacity, power, appears in transformed human acts, shaped by contemplation, human selves lightened, in transfixity, as Cassian says, to a bit of down in a breeze. The crowning theandric act may be to sensuously peer, or notice one's way into, things so that "the logoi [essence, self-likeness] of everything that is divided and particular" (Louth, *Ambiguum* 41, 1313A–B) is interiorly savoured, while all the relations of these things, cliff, grass, ripening plums, their playful unity, how they touch and assist and quicken to resplendence one another is also masticated within. The mind, spreading light, sees light spreadingly implanted in objects of comprehension. Then one is home.

✶ An unhinging time involving mass deaths in the biosphere, like the more than a billion sea shore animals killed during the heat dome on the BC coast in June 2021, is contemporary with the collapse of orienting philosophical and political world systems, this giving birth to rage-filled populisms. There is much inducement to despair. But one must stay at one's post and engage in preserving activisms, which include such epistemological retrievals and re-enactments of ancient forms of knowing and Dionysian pedagogical styles as has been attempted here, checks to what Val Plumwood called "master reason" and theologian Willie Jennings calls "whiteness" as a way of knowing that values command, mastery of material, the essence of the academy. The Dionysian way in Plato, pseudo-Dionysius and Maximus the Confessor tendrils, I like to think, from the Orphic tradition through Pythagoras, which has its own roots in the Upper Paleolithic revolution, the old homo sapiens's school. This synthesis survived through the last ice age; it may make a kind of home for us now.

Going back, the difficult act of home-making, cannot happen without conciliation with Indigenous peoples, and this can start through listening and entering relationships and listening in on first-language revitalizations and on projects to rename colonized places. Mt. Tolmie is SN̲, AK̲E. And going back involves walking into a bewildered and grieving compunction about what we newcomers did in residential schools.

Diotima, Again
Desire is a changeling. The son of Poros and Penia, "tough and shriveled and shoeless and homeless, always lying on the dirt without a bed, sleeping at people's doorsteps" (Plato, *Symposium* 203D), is, as we have previously seen, perpetually in need like his mother, forever feverishly en route like his father. But expose this offspring to beauty, physical, moral, intellectual, political, interior and he moves without a moment of thought, without an instant of doubt or shame, savouring, testing the vector of his passage as he moves in it. His capacity for interior tasting is discerning; he finds his friends and sponsors, and these help

him along. "He is by nature neither immortal nor mortal. But now he springs to life when he gets his way; now he dies" (203E).

Diotima, having disarranged the certainties of the young Socrates, shifts gears in the late stages of her dialectic. "What is the real purpose of love," she asks her student, and he replies he has no idea (206B). "Well, I will tell you: it is giving birth in beauty, whether in body or in soul" (206B). And, she adds, we are all almost always pregnant "both in body and in soul." To this description of eros, of human identity, Socrates replies, "Maybe" (206E). "Love wants to possess the good forever," Diotima adds. Maybe.

Some, however, are "even more pregnant in their souls than in their bodies" (209A), carrying inchoate versions of moderation and justice, among the other virtues, or nascent gifts in philosophy or the arts. These persons search even without knowing they do, for someone beautiful to stand beside, tarry with, and out of this comes the birth in them of "what [they have] been carrying around for ages" (209C). This meeting and birthing is precisely what is happening in the back and forth, the rites of love, of Diotima and Socrates.

She then unspools for her mildly amazed pupil the various realms where "births" can take place from young love to the heights of an engaged metaphysics of ecstasy. And one of these worlds in which love reproduces is that of justice, "the beauty of activities and laws" (210C), customs, a stopping place as one travels up to "the great sea of beauty." In the final stages of her instruction on eros, Diotima presents a vision of erotic becoming as an ornate, spectacular, virtually self-moving vehicle, a great human daemonic contraption for interior movement, coursing in the mind of the recipient—this is the craft that will carry her student to his ends. Launched by the conversation, this power proves to be an extraordinary conveyance, like the boat of the mouth of John Cassian, the boat of the mouth of Christopher Okigbo, of Louise Halfe, transporting you on the wisdom journey, shamanic travel, the just citizen's way.

NUMINOUS
SEDITIONS

The Best of the Centaurs
What speaks the out-of-range, the exceedingly apt strangeness that begins a going home, the inrush of oddness that completes the apperception of the real? Homer in *The Iliad* says the "best of the centaurs" speaks (11.831–32). The infrahuman animal, says Robert Brightman, the animal beneath and within, the prophetic, guiding animal, speaks (163); the angel, bright trickle within cascading light, angel as epiphanous cognitive act, speaks; the feral poet nudges you to true sight, as does the musician and the spiritual guide. You learn over time, at the distant reaches of eros, to be comfortable with these unnoted—and in certain contexts apparently non-existent—bundles of bestirment, learn to be permeable to them and discerning. These powers teach; some speak, while others amass exigence within you.

✳ Chiron, daemonic therianthrope, half man, half horse, overseer of the *paideia* of gods and heroes, teaches pharmacology to the father of the god-like Machaon, who saves the life of Menelaos, Agamemnon's son, who has been wounded by a Trojan arrow.

> He loosened the war-belt, all-gleaming, and beneath it
> the underbelt and belt-guard which bronze-working men toiled to make.
> Then when he saw the wound, where the pointed arrow had entered,
> sucking out the blood, he then expertly sprinkled on soothing
> herbs, which once upon a time Chiron with kind intent gave to his father.
> (Homer 4.215–19)

Chiron, animal-man and god, Zeus's half-brother, instructs Achilles in the playing of the rage-soothing lyre and teaches him the art of the healer (Homer 11.831); he teaches the god Dionysos, then a boy, "bacchic rites and mystic solemnities" (Photius *Biblioteca* 190, in I. Hadot 436). Chiron, says Pindar, is nothing less than a sage (*Pythian* 3:63, in I. Hadot 437); he instructs both Jason and Asclepius; Plato judges him the "wisest" (*Hippias Minor* 371D); he is the preciously elsewhere source of fresh, apprehended pattern.

The centaur lived in the forested mountains of Thessaly, ancient equivalent of the desert of the Christian fathers and mothers, choice location of contemplative practice. Ilsetraut Hadot says Chiron unites "all the traits of an ideal educator in himself," possessing "all the means of favorably influencing both body and soul of his pupils" (437). Athletics trainer, physician, spiritual guide, his chief means of directing the soul was music and the song of verse.

✳ Chiron's nature is not altogether strange to us. We have seen this mixed, preternatural essence much earlier in human history in the ecstatic, bird-headed man before the disemboweled bull in Lascaux cave; in the bison-headed-and-shouldered human in the end chamber at Chauvet cave; in the two dancing "sorcerers," reindeer-manlike beings, at Les Trois-Frères. In the Hohlenstein-Stadl lion-man from over 30,000 years ago in Germany, we have seen that one. Typically, in treatment of Paleolithic parietal art in archeological books, such human-animal figures are identified as shamans engaged in the practice of sympathetic magic. But this identification could well be interpretive overreach: these mixed beings are just as likely to be aiding, formational animals alive and at work in the Aurignacion universe.

✻ Human marriages with birds and animals are not unheard of in North American Indigenous, pre-contact literatures. A man marries a goose in a Haida story, and the goose-woman attempts to live in the human world, but discovers she cannot tolerate the food eaten by her new relatives (Bringhurst 32–44). Nevertheless, during a famine, she makes sure her human in-laws receive sufficient goose food, pine noodles and clover roots, and by this they survive.

The hunter sees the one who will be his wife as a woman, bathing in a pond with her sister, her goose skin discarded at the shore, and he takes her home. The goose-woman eventually returns to her people, tired of human gossip, even though "leaving her husband sickened her heart" (36). Her stricken husband, after a time, follows her. After a long journey, resembling an interior passage or an initiation, marked by the learning of ritual and by animal and spirit assistance, he enters her land where he endures his own version of his wife's food difficulty and is returned to his home place under a raven's wing. While the marriage does not last, the fact remains that the story shows no substantial distance between species; the barrier separating geese and human beings is breachable and love and aid can pass between. Transspecies migration is spiritual work.

A comparable tale comes from the Nett Lake Chippewa of Minnesota (Brightman 163–65). A woman undergoes a fast in the bush where she meets a man who takes her to his lodge; there, she is given food and clothing. She lives happily with the stranger, who provides many baskets, more beautiful apparel, more fish and game, plentiful firewood, and she bears four children with him. It is only much later, when she hears humans walking and talking around her lodge, that she has an inkling her mate is in all probability a beaver. Such stories are presented as myths, cousins to whimsy, by many anthropologists, but they are, to my ear, heightened ontology, tales of encounters with merciful and generous animals, active in a soft border band of being accessible to animals and certain humans; or, better, are accounts of meetings within a great river of élan vital swinging through the present, obvious to many animals but rarely to us. Robert Brightman, in *Grateful Prey*, his study of Rock-Cree human and animal interactions based on research in a remote hamlet in northern Manitoba in the 1970s and 1980s, describes a citizenship in the in-between realm as the world of the "'infrahuman' animal, 'infra' here possessing the sense of 'beneath' or 'within'" (163).

✳ We are going through a cold patch on the coast, and the mountain behind the house is all snow and ice. Birds stir slowly. I leave the hummingbird feeder that has been out since 7:30 a.m. in the house to thaw while I go to the shed in the back to write about Brightman's observations on the beneath and within animal. When I return, climbing the stairs to the back door, the male Anna's rushes toward me at hand height, then whizzes around my head, and I have to call my partner to pass the feeder quickly through the doorway. The bird feeds as I hold with a crooked finger the feeder's hook.

✳ Among the nēhiyaw with whom he lived at Pukatawagan, hunters and trappers, Brightman heard neighbours speak of other animal-like helpers, in particular "an entity called *pawākan*. The *pawākan*, literally, dream image, is an individuated *ahcāk* [soul] being with whom persons understand themselves to experience recurrent communications in dreams" (76–77). The spirit is attentive and sympathetic if its human link behaves appropriately. "Whether or not it is identified as an animal, the *pawākan* may be talked about as essential to foraging success, communicating with its human dependent in dreams... Explaining the concept of *pawākan*, Johnny Bighetty said, 'If you're in any kind of trouble, you dream. You dream things, all kinds of things... animals, trees, stones, ice. If you love...if you do everything it says to do, it'll help you'" (Brightman 77). Here the animal performs a function similar to that of the Socratic daimonium, though it may be more loquacious.

✳ Macarius the Egyptian, a former camel-driver and trader in nitre, was a fourth-century Christian monk, the first, says John Cassian, to find a way to live in the harsh desert of Scetis. Having travelled from Scetis to the mountain of Nitria once, he was asked, in the usual way of the early Egyptian solitaries, to offer an instructional word to the gathered brethren. He began by confessing that he had not yet become a monk himself after many years in his cell, but he claimed that he had seen monks. For years, sitting in his hut in the desert, he had been troubled by the thought that he should journey to the deeper wasteland to see what might be there. Finally, after much struggle, he went, and, after a period of time, came to a wide shallow lake with an island in the middle of it; there the animals came from great distances to drink. "In the midst of these animals I saw two naked men, and my body trembled for I believed they were spirits" (Ward 125–26). The two quickly assured Macarius that they were indeed men, who had come to the remote area from a monastery forty years ago. Macarius asked them how they managed this extreme form of practice. Their life arose, they told him, from the act of giving "up all that is in the world"; in this emptiness the severe conditions did not trouble them. "It is God who has made this way of life" among the animals for them, they said. "We do not freeze in winter, and the summer does us no harm."

✴ Andrew Ahenakew was a nēhiyaw (Cree) Anglican priest from Sandy Lake in north central Saskatchewan in the early to mid years of the twentieth century, part of an extended family that had embraced early the new religion of Christianity. The story of his gradual return to nēhiyaw traditions is told by his wife, Alice Ahenakew, in *âh-âyîtaw isi ê-kî-kisk êyitahkik maskihkiy/They Knew Both Sides of Medicine: Cree Tales of Curing and Cursing*. Alice Ahenakew herself had been orphaned at the age of six and was raised by traditional grandparents at Sturgeon Lake.

In the early years of their marriage, Andrew Ahenakew attended on his own a week-long diocesan meeting at the Pas in neighbouring Manitoba. One night, he was praying for his brother who was seriously ill with cancer. After prayer, he went to bed. "All of a sudden," he later reported to his wife, "it seemed as though I was outside" (63). Immediately he noticed an animal, a polar bear, approaching from some distance: "he seemed to come running in mid-air, but as though in mid-air; approaching, he came to a stop here where I was sitting and simply looked at me."

Then the bear addressed him, "It is from a holy place" that he had come, he told the wide-awake sleeper, and as he said this, he looked above himself. "It is from the holy place that I have been sent hither, and I have come to give you my body," he told the priest, "for since God made the earth, when he made the animals, we, we are still as God made us in the very beginning, we have no sickness in our bodies" (63). The bear had appeared now "for you to use, for you to make medicine there from my body and to doctor people who are sick." The priest in reply just looked at his visitor, recalling "the collar," his clerical collar, and the bear caught his hesitation and said "That is what you are thinking, but, in any case, you will do it nevertheless, that which I have come to tell you," and he showed the priest a sample of the medicine that he must make.

Andrew Ahenakew forgot this visitation for a week after he returned to Sandy Lake, but when he recalled his encounter, his dream, he told his wife about it, and she saw immediately the value of what he had experienced; "you have been given something that is sacred," she told him. He must prepare the first batch of medicine, but she would intercede if he hit an obstacle. They eventually heard of a hunter in Cochin, Saskatchewan, who had recently killed a polar bear, and they travelled there to get the specific parts they needed. After the healing of his younger brother, many others sought Andrew Ahenakew out and gradually he saw the power of nēhiyaw practices, which he kept, along with his priestly duties, for the rest of his life. Both he and his wife did this, Alice Ahenakew concluding "The Priest's Bear Medicine" story by honouring "these three forms of worship—the Anglican liturgy, the Roman Catholic liturgy, and the Cree rites, these are the three for me" (81).

The Angel as Cognitive Act
In the initial moments of the incarnation, an extraordinary presence appeared to men living with their sheep in rough pasture land; this manifestation triggered a violent alteration to their immediate surroundings—"and the glory of the Lord shone around them" (Luke 2:9–10)—and this shift in the air, in the vegetation and stones caused these men to be terrified. The effervescence, the heightened surround, was, however, oddly sympathetic to the shepherds' consternation. The unnamed angel, who appeared out of the spreading light as its centring force, was less a being than an event, an event that speaks, an interior event, yet with an outside face, destabilizing, expanding, a flaring emotional noesis. "Do not be afraid," the light said, the subjectivity of the herdsmen that approached an autonomous voice: "I am bringing you good news…to you is born a saviour, who is the Christ" (Luke 2:9–10).

The shepherds immediately went to see what had been described to them. They had received an alien, yet shockingly intimate, expansion of imagination out of nowhere, and they made themselves, were made, witnesses of an inconceivable ontological shift, of which, stunningly, their witnessing somehow made them co-creators. They capped their experience out on the land with a visit to a nearby barn to gaze on a lit scene, immensely expanded through their inflamed interpretation, and then they disappeared into a world that has been knocked graciously, but not always obviously, askew.

✶ Even in the face of climate change, its heat domes, its massive flooding, its fires, I sense the covert, epiphanous, out-of-the-corner-of-the-eye efficacy of the resuscitation, the disinterring, of the West's deep sapiential past in its Paleolithic, Platonic, Neoplatonic, Christian, Judaic and Islamic forms. All, except of course the first, are inheritors of the erotic Platonic tradition, which itself branches from elements in the protocols of the late European Paleolithic, I sense, its rituals and stories. This daylighting of the submerged river of maieutic ingenuity that has sustained millennia will both enliven us as we face the effects of global warming and help to de-imperialize the still dominant, destructive form in philosophy that subsidizes masterly forms of commerce and technology, resource extraction and reason that helped to land us in our present situation. Thus I re-enter Ibn ʻArabi's *Fusus al-hikam* and Corbin's *Creative Imagination in the Sufism of Ibn ʻArabi*; thus I re-enter Eriugena's *Periphyseon*; again I listen to Hadewijch of Antwerp, to Catherine of Siena, all the while favouring a bolder exploration of the tradition's ecstatic end.

As I bring in the feeder to thaw and take it out again, I am conscious that the angels of the Areopagite await my attention. I have been waiting for an opportunity to re-read *The Celestial Hierarchy* for quite some time. These powers, too, are unnoted, in a winter of their own, a freeze of inattention, yet they are quick, mildly merciful and extremely, gracefully potent.

Early each morning when I leave the house, long before anyone in the city is up, to walk to the shed where I pray and work, I look up at the pre-dawn mountain SN̲, AK̲E, the great bulk of its massed quiet. This impacted, stacked silence is an extraordinary, lavish gift. The mountain's *quies* is richly and ingeniously providential, feast for the interior mouth. The mountain never ceases in its quietness. I say to the mountain's *hesychia* T̲E W̲EN̲ QE N̲ ET TEW̲A (Look at us and have mercy on us). It's still too dark for birds; the hummingbirds will remain in their mysterious torpor for another couple of hours or so.

I am repeatedly struck by how unready we are for the cataclysmic age coming over us. I don't mean unready in an engineering or political sense, though we clearly are unprepared in both these areas. A three-day November deluge has taken out many of the bridges and much of the road bed of the Coquihalla highway in western British Columbia, the coast's chief link to the interior, and, months later, much of the damage remains. And I don't mean we are unprepared in an economic or ecological sense, though we are. At the centre of the self we are not ready, do not have the tools to make ourselves ready, for this new life that shows signs of being harrowing. Our interior unreadiness often finds expression in outpourings of rage.

I read in a spirit of *lectio divina* daily in order to be shaped. On the top of my pile beside my chair near the shed window now are *âh-âyîtaw isi ê-kî-kisk êyitahkik maskihkiy/They Knew Both Sides of Medicine: Cree Tales of Curing and Cursing* and notes from two years ago on Bonaventure's eschatological epistemology in *Itinerarium Mentis in Deum* (*The Mind's Road to God*), where he identifies human imagination with the Logos.

✻ *The Celestial Hierarchy* of pseudo-Dionysius the Areopagite begins with a description of a metaphysical vortex, a rapids of energy, higher travelling to lower, lower then bending back in the bed of pneuma. "Every good endowment and every perfect gift is from above, coming down from the Father of lights," he quotes from the Letter of James (1:17), then adds breath-stoppingly—"But there is something more." This more is a cascade of numinal luminosity through the cosmos in which "each procession of Light spreads itself generously toward us, and, in its power to unify, it stirs us up by lifting us up. It returns us back to the oneness and deifying simplicity of the Father who gathers us in" (Pseudo-Dionysius, *Celestial Hierarchy* 120B–121A). This spreading light, tumbling light, pneumatic, rising, love, sentience, teleology, along with an inexpressible more, has its eddies, which are jarrings in us provoking instantly beneficent alienations from the self that make for unity—with yet another self and the real that this self clearly knows.

How badly we need the assuring insight, this saturation of light in everything, at the centre of Dionysian light theory; it adds a necessary weight to us. Bonaventure will assure us that a shard of this light movement is lodged in us in a particular moment of our consciousness of the world, the light erupting instant, the flash of understanding what is.

✶ There is a juridical nothing that is daily, unrelentingly beneath us; this absence is caught in Sylvia McAdam's, saysewahum's, observation that, given Aboriginal title, given the unparalleled aptness of nēhiyaw laws on Treaty 6 land, given the lie of *terra nullius*, the Canadian government has no jurisdiction here on the land (24–25) where we prepare meals, sleep, listen to music, as house prices rise on property that is not in a deep sense owned by the occupant, who no doubt does not know his land's true name. The pancultural lightheadedness intimation of this airy structure fosters places us within a wider, active, shaping nothing, stripping us to a barebones acedia, its freneticism and bitterness. We do not register that we suffer in this state of affairs. We pelt about in our cars. Anyone with the temerity to point out the embarrassment of our grievous lack of chthonic foundation can draw an incredulous, annihilating anger.

But it must be immediately admitted that our chthonic nothingness is doubled by the absence of a contemplative root in us. We are dangerous in our sentimentality and in our committed defence of structures of distraction, professional sports, popular culture, social media that hide the actual abysmal nature of our lives. All we have is the simple tale of our privilege as a just birth-rite and its newer, shadowy Grimm's story, appearing insistently during the COVID-19 pandemic and the horrendous fire summer of 2021, of that privilege's inadmissible losability.

✶ The typical Judeo-Christian depictions of angels—wings, faces, flaming wheels—strike us as both familiar and archaic. The peculiarity, unbelievability, of these representations, says pseudo-Dionysius, is a providential and unexpectedly perfect response to the imperfection of our minds and the intolerance of this state of mentational affairs on the part of our desires. "These pictures," enormously erroneous, "have to do with beings so simple that we can neither know nor contemplate them" (*Celestial Hierarchy*, 137B) and often so quotidian, I add, we imagine they do not require special naming or representation. But bodiless, story-less, unintelligible, they do serve the preservation of the structure of the cosmos, at least our instrumental role in the identification of this structure, and we yearn for contact with what these powers are and do.

These bursts of essential light marking this penetrating understanding, these flares in mind, sometimes appear on our interior palates as terror, followed by a sense of inconceivable mercy. We require such mercy because "we lack the ability to be raised up to the conceptual contemplations" (140A). Instinctively we seek this exquisite anagogia, calling it meaning, vocation, placedness, durable happiness or even just clarity; we seek this on a minute-by-minute basis in all our acts of knowing. "We need our own upliftings that come naturally to us" and can receive them only through the filters of narrative, metaphoricity, describability. These celestial interventions, translations of incomprehensible light to us, "raise before us the permitted forms of the marvelous and unformed sights" (140A). Then there is the need to conceal and, as it were, muffle the rawly spiritual, the shockingly formational, *disciplina arcani*, so that its power may not be thinned away through generous exposure. Partly this daemonic preservation is achieved through the mild comedy resident in the implausibility of the representation, the eagle, the ox, images, which, while they remain a sort of identification of eruptive visitation are, as well, a deflection and the subsequent creation of a contemplative refuge.

These flashes that give us the truth of what is, non-conventional, prophetic or leading insights, make us, as well, elsewhere, at our unsettled best, unlike, *atopos*. They bring residence in elsewhere—ekstasis—into our nature.

✶ These unlikely identifications are "as if" accounts, concessions to the restrictions and proclivities of the human mind, as I've claimed, not true accounts, not art. The gesture in the images, birds, thrones, meta-human beings, is true—these presences are transfixing, from a distance, upending—while being false. "The Word of God makes use of poetic imagery when discussing these formless intelligences but, as I have said, it does so not for the sake of art, but as a concession to the nature of our mind" (137B). So, if angels are not their traditional images, what are they? No one can say given our range. But they can be described by the cognitive effect they have: they bestir amazement, states of dislocation, liminality, fierce attachment, prophetic vocation—nothing equals their moments. And angels may be further described by their maieutic effect on those "visited" or those encountering them, an apparent intent "to uplift our mind in a manner suitable to our nature" (137B).

"We must lift up the immaterial and steady eyes of our minds to that out-pouring of Light which is so primal, indeed much more so, and which comes from the source of divinity, I mean the Father" (121A–B). The generosity of this light urges it to multiply itself into an infinity of receptacles, blue spruce, homo sapiens's cognitional structure and song.

✽ What is "light" noticed in things, the "essential reality, the quiddity," of light, to use Al-Ghazali's language in *The Ninety-Nine Beautiful Names of God* (37)? How does this light appear in the interior sensorium of the knower readied to see it? It is clarity, beauty, an autonomy within complex relationships to nearby things, a seeming providentiality that may be savoured as mercy. It is self in relationship with oaks, hummingbirds, ferns, snow on Snow Mountain. What is the "light" by which the godedness of things is noted, the facilitating noetic disposition, the "light" in the knower? Look at the shepherds—they, at the point of revelation, are marked by poverty; loneliness; availability, typical on the social margins, to revolutionary shifts in imagination and being. Their condition replicates the kenosis of the Logos. True seeing, knowing, is the mentational, dispositional occupation of the Word in us. In all acts of understanding, of sensorial apprehension, delight before beauty and discerning judgement are "vestiges in which we see our God," writes Bonaventure in *The Mind's Road to God* (II.7); and, on the other side of epiphanous understanding, the side of the probing subject, "our mind, illuminated and flooded by such brilliance [of understanding], unless it is blind, can be led through itself to contemplate the Eternal Light" (III.7), which has many names. All knowing, for Bonaventure, is such a journey of the mind to divinity, the vivification lodged in everything.

The light, shard of spreading light, is embedded in the deer lost in deep snow, a centred quiddity. Another flake of that luminosity, common while being idiosyncratic, is tucked in the complexity of human looking at the doe through the window. The angel is the arc between the individuated gatherings of emanated light and the fire at both ends. The celestial hierarchy, the angelic band of possibility, of being, which is ekstasis in us, is, as all hierarchies, "a sacred order, a state of understanding and an activity approximating as closely as possible the divine" (Pseudo-Dionysius, *Celestial Hierarchy* 164D). "It is forever looking directly at the comeliness of God" (165A); that is, it is a communion with erotic being that further eroticizes us, fans us to beauty, implants a wonderous otherwise and elsewhere in us. The world becomes merciful, attentive, generous when you feed yourself to the practice of noticing, of being drawn to essential light in things, receiving their conviviality, of having faith in what you then see. What is wild speaks, what is outside speaks, voice angled, under certain ascetical conditions, to our release and benefit. Here, in taking this up, lies courage. This dwelling is the band of the angel. This is the political spectacle of contemplative virtues surfacing the world as communion. Such behaviour is our deepest loyalty.

✳ My mother died several weeks after her ninety-sixth birthday. I had been lucky enough to catch a plane out of Victoria to be with her in her extended care home in Alberta in her last night hours. Two weeks later I was on the trail to Mystic Beach on the Juan de Fuca strait. I was there because I was trying to gather myself, place or fix myself somehow in this new motherless terrain—she had been the first of the small number of people I'd really talked to in my life. When I made it to the beach, the winter tide was well in, so as I walked along, I was brushing against bushes that began the line between beach and forest. Suddenly, a small bird, a winter wren, flew from some salal at a slant just in front of me and stopped me in my tracks. The bird flew into a crudely stacked rock chimney at a fire pit someone had made at the back of the beach. Strange behaviour for a furtive forest bird, always feeding, I thought, so I kept looking at the hole into which it had vanished. Soon the wren emerged and hopped on stones toward me, and then it stopped and looked at me as I stared at it. Then it hopped closer and looked again at me hard, the long last look of the dear, dear dead, an instant of the larger, unsectioned community, and flew off.

SN̲, AK̲E
January 2022

DREAM
CODA

April 1, 2018, Easter Sunday

I dreamt last night of a large tree that had died and fallen. The reason for this is that it had been a tree caribou rubbed against, and, because the herd had vanished, the tree lacked the oil or chemical element the caribou had transferred to it and had died. Its bark appeared to question where the animals had gone: there were marks on the bark that appeared to be anxious question marks.

February 2, 2020

A dream from early this morning. The head and neck of a large doe stick out of our kitchen cupboard. I am amazed to see this and begin to stroke her neck. She is alert, trepidatious, but strangely also calm. As I examine the cupboards it becomes clear that her whole body is mostly in there, with part of it extending beyond the cupboard's capacity; this part of her reclining body is supported by a taped cardboard structure. A hole has been cut in the cupboard door for the deer's head, and it becomes clear that she must be fed. Then, suddenly, at the south end of the kitchen, the doe's fawn (well grown) turns up, walking freely, with a leather collar around her neck. I sense that Helen has come up with the collar. The fawn must be fed as well and I search for food for her. The animals are in the kitchen because some trauma threatens them outdoors, but as I lead the fawn around, I realize this is an opportunity to establish a close bond. All of us will return to the mountain's foot, its ferns and grasses.

March 9, 2020
Another strange dream of the wild entwined with the house. I go outside and I see a large tree growing from the roof of the house. It strikes me immediately it is an invasive plant and must be taken out; someone, me, I suspect, has sawn off its end, which before would have trailed on the ground. Its bark and curves are those of an arbutus. I think later, still in bed, of the tree that grows through the bedroom of Penelope and Odysseus. I hope I do not damage the tree further.

GLOSSARY

Apokatastasis
It was a medical term in the ancient world naming the restoration to an original state, as in the resetting of a bone. In metaphysics and psychology, it is the return to paradise—"to restitute, restore," "to set in order" (*Klein's Comprehensive Etymological Dictionary*).

Catanyxis
It is a possible first step in dispositional transformation, "a sudden shock," Irénée Hausherr tells us, "an emotion which plants deep in the soul a feeling, an attitude" of compunction. It is a "piercing" that "comes from without" (Hausherr 8–9).

Haecceity
The word comes from *haecceitas*, this or thisness. It identifies the aspect of a person, object, place that makes them singular, unlike anything else. In the philosophy of John Duns Scotus, it is the highest form of something, the aching truth of an object or person.

Hesychia
This is stillness that marks the emergence of the true self, the essential self, in the contemplative philosophy of Evagrius and elsewhere in ancient monasticism. Jean Leclercq, OSB, describes this state as that calm, security, "that leisure, that reality so rich that it cannot be circumscribed by any words" (xii) issuing in a life centred on contemplative attention.

Penthos

Irénée Hausherr identifies this word and state as "a sorrowful disposition of the soul caused by the privation of something desirable," quoting Gregory of Nyssa. The tears of this sadness are seen as the term of the erotic life. Hausherr adds to the meaning of the word by writing that "[p]enthos without thanksgiving would be despair," while thanksgiving without *penthic* sadness "would be presumptuous illusion" (Hausherr 18–19).

Phronesis

Wisdom, discernment, the ability to distinguish the true from the inexact or fraudulent. To be "sane, sensible, prudent," "to have understanding, be wise" (*Klein's Comprehensive Etymological Dictionary*).

Stabilitas

This disposition is a willed and sought stability, one of the Benedictine vows (St. Benedict of Nursia Prologue, par. 8; ch. 58), a rooting and taking nourishment from one place, which becomes aspects of oneself.

Statio

"A standing, a standing still," "a place of abode, a resting place" (*Cassel's Latin Dictionary*). In statio, you stand interiorly motionless, gathering attention, seeping into what you see or the task before you.

Ta'wil

Ta'wil is interpretation premised on the sense that texts, acts and things are composed of both exoteric and esoteric elements. *Ta'wil* is a deeper reading, an energetic exegesis, a "prophetic hermeneutic" (Corbin, *Creative Imagination* 90) that leads to a theophanic realism. It can be stirred by catastrophe, among other things.

Theurgy

It is a form of knowing that can issue from ceremony, conversation, striking poetry that may result in the transformation of the self. The word means "divine work."

READING

Addas, Claude. *Quest for the Red Sulphur: The Life of Ibn 'Arabi*. Translated by Peter Kingsley, The Islamic Text Society, 1993.

Adorno, Theodor. "The Essay As Form." *The Adorno Reader*, edited by Brian O'Connor, Blackwell, 2000, pp. 91–111.

Ahenakew, Alice. *âh-âyîtaw isi ê-kî-kiskêyihtahkik maskihkiy/They Knew Both Sides of Medicine: Cree Tales of Curing and Cursing*. Edited by W.C. Wolfart and Freda Ahenakew, U of Manitoba P, 2000.

Al-Ghazali. *The Ninety-Nine Beautiful Names of God*. Translated by David B. Burrell and Nazih Daher, Cambridge UP, 1992.

Alighieri, Dante. *The Divine Comedy*. Translated by John Ciardi, W.W. Norton, 1977.

Arendt, Hannah. *The Origins of Totalitarianism*, 1951. Harcourt, Brace and World, 1973.

Athanassiadi, Polymnia. Introduction. Damascius, pp. 9–73.

Augustine of Hippo. *Confessions*. Translated by Henry Chadwick, Oxford UP, 1992.

St. Benedict of Nursia. *The Rule of St. Benedict*. Edited by Timothy Fry, OSB, The Liturgical Press, 1981.

Berkovits, Eliezer. *Faith after the Holocaust*. KTAV Publishing House, 1973.

St. Bonaventure. *The Mind's Road to God (Itinerarium Mentis in Deum)*. Translated by George Boas, Merrill Company, 1953.

St. Bonaventure. *The Soul's Journey to God, The Tree of Life, The Life of St. Francis*. Translated by Ewert Cousins, Paulist Press, 1978.

Borrows, John. *Law's Indigenous Ethics*. U of Toronto P, 2019.

Brightman, Robert. *Grateful Prey: Rock-Cree Human-Animal Relationships*. Canadian Plains Research Center, U of Regina P, 2002.

Bringhurst, Robert. *A Story as Sharp as a Knife: The Classical Haida Myth-tellers and Their World*. Douglas and McIntyre, 1999.

Carlson, Keith Thor. *The Power of Place and the Problem of Time: Aboriginal Identity and Historical Consciousness in the Cauldron of Colonialism*. U of Toronto P, 2011.

Cassian, John. *The Institutes*. Translated by Boniface Ramsey, Paulist Press, 2000.

Chittick, William C. *The Sufi Path of Knowledge: Ibn al-'Arabi's Metaphysics of Imagination*. State U of New York P, 1989.

Coleridge, Samuel Taylor. *Biographia Literaria*. Edited by Adam Roberts, Edinburgh UP, 2014.

Corbin, Henry. *Avicenna and the Visionary Recital*. Translated by Willard R. Trask, Routledge and Kegan Paul, 1960.

Corbin, Henry. *History of Islamic Philosophy*. Translated by Philip Sherrard, Kegan Paul, 1993.

Corbin, Henry. *Creative Imagination in the Sufism of Ibn 'Arabi*. Translated by Ralph Manheim, Princeton UP, 1997.

Couture, Joseph. *A Metaphoric Mind: Selected Writings of Joseph Couture*. Athabasca UP, 2013.

Cunsolo, Ashlee. "Climate Change As the Work of Mourning." *Mourning Nature: Hope at the Heart of Ecological Loss and Grief*, edited by Ashlee Cunsolo and Karen Landman, McGill-Queen's UP, pp. 169–89.

Cunsolo, Ashlee, and Karen Landman, "Introduction: To Mourn Beyond the Human." *Mourning Nature: Hope at the Heart of Ecological Loss and Grief*, edited by Ashlee Cunsolo and Karen Landman, McGill-Queen's UP, pp. 3–26.

Dagli, Caner K. Introduction. Ibn 'Arabi, pp. xv–xxxi.

Damascius. *The Philosophical History*. Translated by Polymnia Athanassiadi, Apamea Cultural Association, 1999.

Day, Dorothy. "On Pilgrimage." *The Catholic Worker*, February 1, 1969.

Dickinson, Mark. *Canadian Primal: Poets, Place and the Music of Meaning*. McGill-Queens UP, 2021.

Domanski, Don. *Poetry and the Sacred*. Institute for Coastal Research, 2006.

Domanski, Don. *Earthly Pages: The Poetry of Don Domanski*. Wilfrid Laurier UP, 2007.

Domanski, Don. *Selected Poems, 1975–2021*. Corbel Stone Press, 2021.

Eriugena, John Scotus. *Periphyseon*. Translated by I.P. Sheldon-Williams and John J. O'Meara, Bellarmin/Dumbarton Oaks, 1987.

Eshleman, Clayton. *Juniper Fuse: Upper Paleolithic Imagination and the Construction of the Underworld*. Wesleyan Press, 2003.

Evagrius Ponticus. *Praktikos and Chapters on Prayer*. Translated by John Eudes Bamberger, Cistercian Publications, 1981.

Fackenheim, Emil. *The Religious Dimension in Hegel's Thought*. Beacon Press, 1967.

Fackenheim, Emil. *God's Presence in History*. New York UP, 1970.

Forest, Jim. *All Is Grace: A Biography of Dorothy Day*. Orbis Books, 2011.

Grant, George. "In Defence of North America." *Technology and Empire*, House of Anansi Press, 1969, pp. 3–30.

Grayson, Donald. *Thomas Merton and the Noonday Demon*. Cascade Books, 2015.

Gregory of Nyssa. *The Life of Moses*. Translated by Abraham J. Malherbe and Everett Ferguson, Paulist Press, 1978.

Hadot, Ilsetraut. "The Spiritual Guide." *Classical Mediterranean Spirituality: Egyptian, Greek, Roman*, edited by A.H. Armstrong, Routledge and Kegan Paul, 1986, pp. 436–459.

Hadot, Pierre. *Philosophy as a Way of Life: Spiritual Exercises from Socrates to Foucault*. Translated by Michael Chase, Blackwell Publishers, 1995.

Halfe, Louise. *Blue Marrow*. Coteau Books, 2004.

Halfe, Louise. *The Crooked Good*. Coteau Books, 2009.

Harrison, Roberto. *Yaviza*. Atelos 41, 2017.

Harrison, Roberto. *Tropical Lung: exi(s)t(s)*. Omnidawn, 2021.

Hausherr, Irénée. *Penthos: The Doctrine of Compunction in the Christian East*. Cistercian Publications, 1982.

Heidegger, Martin. *Basic Writings*. Translated by David Farrell Krell, Harper Collins, 1992.

Hegel, G.W.F. *Faith and Knowledge*. Translated by H.S. Harris, State U of New York, 1977.

Hegel, G.W.F. *The Phenomenology of Spirit*. Translated by A.V. Miller, Oxford UP, 1979.

Hellner-Eshed, Melila. *A River Flows from Eden: The Language of Mystical Experience in the Zohar*. Stanford UP, 2009.

Homer. *The Iliad*. Translated by Caroline Alexander, Harper Collins, 2015.

Iamblichus. *On the Mysteries*. Translated by Emma C. Clarke, John D. Dillon, and Jackson P. Hershbell, Society of Biblical Literature, 2003.

Ibn 'Arabi. *Fusus al-hikam*. Translated by Caner K. Dagli, Great Books of the Islamic World, 2004.

Izutsu, Toshihiko. *Sufism and Taoism: A Comparative Study of Key Philosophical Concepts*. U of California P, 2016.

Julian of Norwich. *Revelations of Divine Love*. Translated (into modern English) by Clifton Wolters, Penguin Books, 1966.

Karnes, Michelle. *Imagination, Meditation and Cognition in the Middle Ages*. U of Chicago P, 2011.

Lao-tzu. *Tao Te Ching*. Translated by Stephen Addiss and Stanley Lombardo, Hackett Publishing Company, 1993.

Leclercq, Jean (OSB). Preface. Evagrius Ponticus, pp. 1–16.

Lewis-Williams, David. *The Mind in the Cave: Consciousness and the Origins of Art*. Thames and Hudson, 2002.

Lewis-Williams, David, and David Pearce. *Inside the Neolithic Mind: Consciousness, Cosmos and the Realm of the Gods*. Thames and Hudson, 2005.

Lilburn, Tim. *Kill-site*. McClelland and Stewart, 2003.

Lilburn, Tim. *The House of Charlemagne*. U of Regina P, Oskana Poetry and Poetics, 2018.

Lilburn, Tim. "Hoping for Something to Appear." *Selected Poems, 1975–2021*, by Don Domanski, Corbel Stone Press, 2021.

Louth, Andrew. *Maximus the Confessor*. Routledge, 1996.

Luxemburg, Rosa. *Rosa Luxemburg Speaks*. Edited by Mary Alice Waters, Pathfinder Press, 1970.

Mandelstam, Osip. *Selected Essays*. Translated by Sidney Monas, U of Texas P, 1977.

Marion, Jean-Luc. *In the Self's Place: The Approach of St. Augustine*. Translated by Jeffery L. Koskey, Stanford UP, 2012.

Maritain, Jacques. *Creative Intuition in Art and Poetry*. Pantheon Books, 1953.

Marx, Karl. *The Manifesto of the Communist Party. The Marx-Engels Reader*, 2nd ed., edited by Robert C. Tucker, W.W. Norton and Company, 1978.

Maybaum, Ignaz. *An Ignaz Maybaum Reader*. Edited by Nicolas de Lange, Berghahn Books, 2001.

McAdam, Sylvia (saysewahum). *Nationhood Interrupted: Revitalizing* nêhiyaw *Legal Systems*. Purich Publishing, 2015.

Merton, Thomas. *Monastic Observances*. Edited by Patrick F. O'Connell, Cistercian Publications, 2010.

Norris, Kathleen. *Acedia and Me: A Marriage, Monks and a Writer's Life*. Riverhead Books, 2008.

Nussbaum, Martha. *Love's Knowledge: Essays on Philosophy and Literature*. Oxford UP, 1992.

Nussbaum, Martha. *The Therapy of Desire: Theory and Practice in Hellenistic Ethics*. Princeton UP, 2018.

Palladius. *The Lausiac History*. Translated by Robert T. Meyer, Newman, 1965.

Paul, Philip Kevin. *Taking the Names Down from the Hill*. Nightwood Editions, 2003.

Paul, Philip Kevin. *Little Hunger*. Nightwood Editions, 2008.

Plato. *Apology of Socrates. Four Texts on Socrates: Plato's Euthyphro, Apology, and Crito and Aristophanes' Clouds*. Translated by Thomas G. West and Grace Starry West. Cornell UP, 1984, pp. 63–97.

Plato. *Parmenides*. Translated by R.E. Allen, U of Minnesota P, 1983.

Plato. *Symposium*. Translated by Alexander Nehamas and Paul Woodruff, Hackett Publishing, 1989.

Plato. *Republic*. Translated by Allan Bloom, Basic Books, 1991.

Plato. *Ion, Hippias Minor, Laches, Protagoras*. Translated by R.E. Allen, Yale UP, 1996.

Poitras, Edward, and Robin Poitras. "House of Chow Mein." Performance, U of Regina, September 2015.

Proclus. *Proclus' Commentary on Plato's Parmenides*. Translated by Glenn R. Morrow and John M. Dillon, Princeton UP, 1987.

Pseudo-Dionysius the Areopagite. *The Celestial Hierarchy. The Complete Works*. Translated by Colm Luibheid, Paulist Press, 1987, pp. 145–91.

Pseudo-Dionysius the Areopagite. *The Ecclesiastical Hierarchy. The Complete Works*, pp. 193–259.

Riel, Louis. *The Collected Writings of Louis Riel*. Edited by George G.F. Stanley, Raymond J.A. Huel, Gilles Martel, Thomas Flanagan, and Glen Campbell, U of Alberta P, 1985.

Robertson, Duncan. *Lectio Divina: The Medieval Experience of Reading*. Liturgical Press, 2011.

Rose, Gillian. *The Broken Middle: Out of Our Ancient Society*. Blackwell, 1992.

Rose, Gillian. *Mourning Becomes the Law*. Cambridge UP, 1996.

Rubenstein, Richard. *After Auschwitz*. Macmillan, 1966.

Rustom, Mohammed. *Qur'anic Exegesis in the Later Islamic Philosophy: Mulla Sadra's Tafsir Surat al-Fatiha*. 2009. U of Toronto, PhD dissertation.

Rustom, Mohammed. *The Triumph of Mercy: Philosophy and Scriptures in Mulla Sadra*. State U of New York P, 2012.

Shanks, Andrew. *Against Innocence: Gillian Rose's Reception and Gift of Faith*. SCM, 2008.

Shaw, Gregory. *Theurgy and the Soul: The Neoplatonism of Iamblichus*. Pennsylvania UP, 1996.

Szumigalski, Anne. *On Glassy Wings: Poems New and Selected*. Coteau Books, 1997.

Thoreau, Henry David. *Walden*. Beacon Press, 2004.

Thubron, Colin. "Cartographers of Stone and Air." *The New York Review of Books*, December 3, 2020.

Treaty 7 Elders and Tribal Council, with Walter Hildebrandt, Dorothy First Rider, and Sarah Carter. *The True Spirit and Original Intent of Treaty 7*. McGill-Queen's UP, 1996.

Walker, Michelle Boulous. *Slow Philosophy: Reading Against the Institution*. Bloomsbury, 2017.

Ward, Benedicta. *The Sayings of the Desert Fathers: The Alphabetical Collection*. Cistercian Publications, 1984.

Weil, Simone. "God in Plato." *Simone Weil, Late Philosophical Writings*, translated by Eric O. Springsted and Laurence E. Schmidt, Notre Dame UP, 2015.

Zwicky, Jan. *Auden As Philosopher: How Poets Think*. Institute for Coastal Research, 2011.

INDEX

Abraham, 20
Academy (Athens), 122, 123, 141
acedia, 93–95, 96, 97, 100, 101–02, 155
Adam and Eve, 107
Adorno, Theodor
    "The Essay As Form," 30
agent intellect (active intelligence)
    Bonaventure on, 21, 22–23, 34, 97, 101
    as visitation of the angel, 19–20, 22–23, 34–35, 98, 127, 138, 139
agronomic universities, 9, 13
Ahenakew, Andrew and Alice, 151–52
Alcibiades, 104, 139
alētheia, 67, 85
Al-Ghazali, 157
Ammons, A.R., 4
angels
    agent intellect as visitation of, 19–20, 22–23, 34–35, 98, 127, 138, 139
    as cognitive act, 147, 152–53, 156–58
    depictions of, 155–56
anger, 92–93, 95
animals, infrahuman, 147–52
Anselm of Canterbury, 32, 51
Anthony the Great, 96, 97
apokatastasis, 46, 92
Arendt, Hannah, 115–16
Aristophanes, 104–05, 106

Aristotle, 34
art and creativity, 3–4, 34, 47
ascesis, 9–10, 96, 136–37, 138, 141, 143
Asmus, Rudolf, 125
Athanassiadi, Polymnia, 121, 124–25
atopos, 50, 62, 156
Auden, W.H., 33, 34
Augustine of Hippo, 77–78, 79–80, 82–84, 87
autochthonicity, 2, 10, 19, 44–45
Avicenna, 27, 29, 34, 78, 87–89
    The Recital of the Bird, 88–89
    'ayan (ultimate identity of a thing), 84–86

beauty, 20, 21. *See also* eros
Benedictine tradition, 17, 102, 110–11
Benedict of Nursia
    The Rule of Benedict, 42
Berkovits, Eliezer, 7
Bernard of Clairvaux, 2, 22
bewilderment, 68–69, 77–78, 81, 84
Bighetty, Johnny, 150
Bishop, Elizabeth, 1
Blake, William, 31, 40, 72
Boehme, Jacob, 72
Bonaventure
    on agent intellect, 20–21, 22–23, 34

169

Itinerarium Mentis in Deum (*The
  Mind's Road to God*), 21, 97, 154,
  157
Life of Francis, 97–99, 100–01
on light of true knowing, 155, 157
Borrows, John, 45
Brightman, Robert, 147, 149–50
Burtynsky, Edward, 4

capitalism, 115, 137
Carlson, Keith Thor, 10
catanyxis, 4, 9, 26, 39, 46, 124
catastrophe
  Catholic Worker's approach to, 8–10
  climate change and, 2, 5, 7–8
  Fackenheim's approach to Holocaust,
    6–8, 10
  language and, 92
  Stó:lō approach to, 10
Catherine of Siena, 153
Catholic Church. *See* Christianity;
  Roman Catholic Church
Catholic Worker movement, 8–10
ceremony, 138–39
Chiron, 147–48
Chittick, William, 65, 66, 68, 71, 78
Christianity, 122, 140–42, 143–44. *See
  also* Roman Catholic Church
Cistercian tradition, 110–11
civil society, 116–17. *See also* politics
climate change
  acedia and, 97
  action, need for, 42
  autochthonicity and, 2
  as catastrophe, 2, 5, 7–8
  common responses to, 6, 8, 16, 131
  epistemology and, 132
  heat dome (2021), 136, 145
  interiority and, 2, 4–6, 7–8, 9–10,
    12–13, 16–17
  mourning what is lost, 17–18
  religion and, 132–33
  unpreparedness for, 154

Coleridge, Samuel Taylor, 33–34
colonialism, settler, 2, 44–45, 76,
  132, 133, 137, 141, 155. *See also*
  imperialism
confession, 38–39, 78, 82, 103
confusion. *See* bewilderment
contemplation
  approach to, 29–30
  absence of, 155
  ceremony and, 138–39
  goals of, 18–19
  knowing and, 33–36, 157–58
  lectio divina, 13, 27, 30–33, 35, 41,
    111, 120, 154
  Maximus the Confessor and, 137–38,
    142–43
  Middlebury College's "mindfulness
    project," 27–28
  place and, 44–47
  poetry, philosophy, and
    contemplative attention, 57,
    59–60, 62–63
  sharing the fruits of one's
    contemplation, 36–42
  teaching and, 25–27, 28–29
  theandric energy and, 144–45
  See also interiority; self
conversatio morum, 42, 95, 138
conversation, philosophical, 12–13,
  36–37, 39, 41
Corbin, Henry, 40, 66, 71–72, 78, 85, 87,
  142, 153
Creates, Marlene, 4
creativity and art, 3–4, 34, 47
Cree (nēhiyaw), 30, 150, 151–52, 155
Crozier, Lorna, 34
Cunsolo, Ashlee, 45

Dagli, Caner K., 84–85, 86
Damascius, 120–29
  background, 122–23
  contemplative pedagogies and,
    26–27

contemporary experience of reading, 120–22
digression and, 124–25
on Hypatia and Hierocles, 128–29
Isidore and, 122, 124, 125, 126–27
monastic exempla and, 127–28
The Philosophical History, 121, 123, 124, 125
philosophical inquiry and, 114, 125–26, 129
Day, Dorothy, 6, 8–10, 11, 13
depression, 5, 94. *See also* acedia
Depression, Great, 6, 8–9
deracination, 44, 106–07
Descartes, René, 26, 80, 106
"The Descent of Inanna," 57, 59
desire, 13, 42–43. *See also* eros; love
Dewdney, Christopher, 4
dialectic, 12, 39, 40, 125
Di Cicco, Pier Giorgio, 59
Dickinson, Mark
  Canadian Primal, 44
digression, 124–25
Dionysian tradition
  Christianity and, 140–42
  contemporary relevance of, 132, 142, 145
  Diotima on love and, 134–35, 145–46
  Maximus the Confessor and, 136–38, 142–43, 144–46
  pseudo-Dionysius and, 138–40
Diotima
  about, 132
  comparison to Avicenna's *Recital of the Bird*, 88–89
  instructional style, 133–35
  on love, 105, 134–35, 145–46
  pseudo-Dionysius and, 139
  Socrates and, 29, 140, 141, 142
discernment, 12, 32, 85, 95–96, 127, 134, 142–43
disciplina arcani, 38, 67, 125, 156

Domanski, Don, 49–53
  "A Thin Place," 51, 52
  "Edge," 49
  "Nocturne," 53
  "One for an Apparition," 50
  "Osprey and Salmon," 51
  "Poetry and the Sacred," 52
  "Small Hours," 53–54
  "Wolf-Ladder," 51
Duns Scotus, John, 108

ecstasy, 13–15, 21, 87. *See also* eros; love; visionary recitals
Eliot, George
  Middlemarch, 31
environmental disasters, 109, 110. *See also* climate change
epektasis, 59
epiphany, 26, 99, 103
epistemology, 4, 67–68, 86–89, 101, 114, 132, 138, 141
Eriugena, John Scotus, 26, 136
  Periphyseon, 153
eros
  Julian of Norwich and, 57
  philosophy and, 59, 61–62
  practice of, 78, 82–84, 158
  Rose and, 119
  sharing the fruits of one's contemplation and, 36, 39–40
  See also desire; love
Eshleman, Clayton, 4
esoteric, 75–76, 140–41. *See also* Dionysian tradition
ethics, 61, 113
Evagrius Ponticus, 27, 30, 91, 92–93, 94–96
example
  Damascius's use of monastic exempla, 127–28
  sharing by way of, 41

Fackenheim, Emil, 6–8, 9, 10, 11

finding (tasting), 68–69, 80–81, 87
Franciscan tradition, 100–01, 102
Francis of Assisi, 21, 97–99, 101, 102
free fall, 35, 36, 42
Freud, Sigmund, 33, 34

gardeners, 12
Gertrude of Helfta (Gertrude the Great)
    The Herald of God's Loving-Kindness, 38, 41
Ghandl (Haida poet), 57
Glaucon, 58
global warming. *See* climate change
Grant, George, 44–45
Grayson, Donald, 94, 100
Great Depression, 6, 8–9
Gregory of Nyssa, 59, 95
Gregory the Theologian (Gregory of Nazianzen), 136

Hadewijch of Antwerp, 153
Hadot, Ilsetraut, 148
Hadot, Pierre, 29
haecceity, 3, 21, 81, 82, 86, 108, 111
Haida, 57, 149
Halfe, Louise, 146
Hall, Donald, 53
Hardy, Thomas
    Tess of the d'Urbervilles, 31
Harrison, Roberto
    Yaviza, 15–16
Hartman, Geoffrey, 119
heat dome (2021), 136, 145
Hegel, G.W.F., 43, 114, 117, 118–19, 124, 126
    The Phenomenology of Spirit, 43, 114, 118–19
Heidegger, Martin, 56, 61, 62, 67–68, 117
Heiti, Warren, 104–05
Hellner-Eshed, Melila, 14–15
Heraclitus, 46
hesychia, 91, 94, 102

Hierocles, 128–29
Hillman, Brenda, 44
Holocaust, 6–7, 61
Homer, 57
    The Iliad, 58, 147, 148
hope, 6, 12
Hsieh Ling-yun, 40
Hugh of St. Victor, 31–32
human-earth severing, 106–08
Hu Zhu Dong, 1
Hypatia, 128

Iamblichus, 26–27, 46, 123, 126–27
Ibn 'Arabi
    angelic visitations and, 38, 126
    biography, 66
    eros, practice of, and, 83
    Fusus al-hikam (*The Ringstones of Wisdom*), 41, 71, 85–86, 153
    Futuhat al-makkiyya (*The Meccan Openings*), 38, 67, 81, 84, 86–87, 126
    on interior formation, 20
    mundus imaginalis and, 40–41, 71
    reading through Chittick, 65, 66, 68, 71–72
    on Reality's "self-disclosure," 67–68, 80–81, 86–87
    on seeing things as they really are, 84–86
    on ungraspability of self, 77–78, 79, 81–82
Ignatius of Loyola, 13, 29
imagination, 13–17, 22, 33–34, 36, 40–41, 53, 81–82, 143–44
imperialism, 107. *See also* settler colonialism
Indigenous Peoples
    author learning from, 30
    colonial treaty-making and, 76
    Dionysian tradition and, 142
    Haida, 57, 149
    nēhiyaw (Cree), 30, 150, 151–52, 155

Nett Lake Chippewa, 149
reconciliation, 45–46, 145
residential schools, 131–32, 133, 145
SENĆOŦEN language, 1, 2, 132
Stó:lō, 10
stories of human marriages with animals, 149
W̱SÁNEĆ, 13, 17–18, 30, 40, 44, 47, 91–92
infrahuman animals, 147–52
innocence, 116
intelligence. *See* agent intellect
interiority
  about, 11
  agent intellect (active intelligence) and, 19–21, 22–23, 34–35, 97, 101, 127, 138, 139
  climate change and, 2, 4–6, 7–8, 9–10, 12–13, 16–17
  and conversation and spiritual direction, 12–13, 20
  Day and Maurin's approach, 8–10
  distractions from, 92–93
  Fackenheim's approach, 6–8, 10
  hesychia and, 91, 94, 102
  imagination and, 13–17
  somatic repetition and, 110–12
  unpreparedness for, 154
  See also acedia; contemplation; self
Isidore, 37, 122, 124, 125, 126–27
Izutsu, Toshihiko, 71

Jennings, Willie, 145
Jesus Christ, 86, 122, 143–44
John, Gospel of, 86
John Cassian, 34, 93, 144, 146, 150
John Chrysostom, 39
John of the Cross, 40, 126, 140–41
John Paul II (pope), 142
Julian of Norwich, 57
justice, 18–19, 28, 46–47, 146

Kant, Immanuel, 106

Keats, John, 35
knowing, contemplative, 33–36

land, 44–47
language, 18, 91–92, 109
Last Mountain Lake (SK), 65–66
Leclercq, Jean, 91
lectio divina, 13, 27, 30–33, 35, 41, 111, 120, 154
Leopold, Aldo, 44
Levertov, Denise, 44
Levinas, Emmanuel, 61
light, 145, 152–53, 154–55, 156, 157–58
Lilburn, Tim
  Domanski, relationship with, 50, 52, 53
  dreams, 159–60
  at Last Mountain Lake, 65–66
  mother, 66, 67, 72–75, 76, 158
Lilburn, Tim, works
  "Cranes, Last Mountain Lake," 69–71
  The House of Charlemagne, 14, 66
  "Kill-site," 60–61
  "The Pavilion, the Veranda Circling, Hanging kerosene Lamps," 72–75
loneliness, ontological, 44, 104, 106–09
love
  Aristophanes on, 104–06
  Day on, 9
  Diotima on, 105, 134–35, 145–46
  pseudo-Dionysius on, 139
  See also desire; eros
Luxemburg, Rosa, 115, 120

Macarius the Egyptian (Macarius of Scete), 95–96, 150–51
maieutics, 4, 11, 19, 26, 28–29, 157
Mandelstam, Osip, 34, 137–38
Marion, Jean-Luc, 78, 79
Maritain, Jacques, 52
Marx, Karl, 44, 106–07
match-making, 26, 39

Maurin, Peter, 6, 8–10, 13
Maximus the Confessor
    about, 30, 132
    Ambigua, 136
    Ambiguum 5, on theandrism, 143–45
    Ambiguum 10, on contemplation and discernment, 136–37, 142–43
    Dionysian tradition and, 145
    epistemology of, 137–38
Maybaum, Ignaz, 5, 6
McAdam, Sylvia, 155
mercy, works of, 9, 13
Merton, Thomas, 99–100, 110–11, 142
metaphor, 3, 4, 108
Middlebury College, 27–28
midrashic tradition, 7, 119
Mill, J.S.
    On Liberty, 44
mimesis, 4
monastic exempla, 127–28
monasticism, 17, 42, 96, 109, 110–11
mourning, 17–18, 45, 97, 117, 120. See also *penthos*; sadness
Muir, John, 44
Mullah Sadra, 111
mundus imaginalis (imaginal world), 40–41, 66, 71, 76
Murdoch, Iris, 57
mystical theology, 61–62, 81. See also Dionysian tradition

negative capability, 35
nēhiyaw (Cree), 30, 150, 151–52, 155
neoliberalism, 107, 115
Neruda, Pablo, 26
Nett Lake Chippewa, 149
Norris, Kathleen, 94
nostalgia, 5, 93, 110
nothingness, chthonic, 155
Nussbaum, Martha, 30, 40

Okigbo, Christopher, 146

O'Leary, Peter, 40
    Earth Is Best, 15
ontology, 61
outside, being on the, 1–2

Palladius
    The Lausaic History, 95
paradoxa, 123, 124–25, 126
Parmenides, 29, 37, 134
patria, 121, 123, 124–25
Paul, Philip Kevin, 13
pawākan, 150
peace
    Francis's "ecstatic peace," 20, 21, 23, 97
    as *hesychia*, 91, 94, 102
pedagogies, contemplative. See contemplation
penthos, 5, 109, 117, 119. See also mourning; sadness
philosophy
    "broken middle," 43, 114, 115–16, 119
    contemplative attention and, 57, 59, 62
    contemporary dismantling of, 113–14, 114–15
    conversation, philosophical, 12–13, 36–37, 39, 41
    Damascius on philosophical inquiry, 114, 125–28, 129
    Diotima's instructional style, 133–35
    epistemology, 4, 67–68, 86–89, 101, 114, 132, 138, 141
    erotics as first philosophy, 61–62
    ontology, 61
    paradoxa and, 126
    Plato on, 56, 57–59
    poetry and, 56–57
    "slow philosophy" movement, 30
phronesis, 134, 143
Pindar, 148
place, 44–47

Plato
    Allegory of the Cave, 14, 15, 88
    on *atopos*, 50
    on Chiron, 148
    on dialectic, 12
    Dionysian tradition and, 145
    on Diotima on love, 105, 134–35, 145–46 (*see also* Diotima)
    interior formation and, 19
    on match-making, 26
    on philosophical wisdom, 114
    on poetry and philosophy, 56, 57–59
    Socrates and, 125 (*see also* Socrates)
    on split self, 43
Plato, works
    Parmenides, 36–37, 134
    Republic, 12, 15, 19, 57–59
    Symposium, 43, 104–06, 133–35, 145–46
Platonisms, 26–27, 46, 141
Plotinus, 26
Plumwood, Val, 145
poetry
    as antidote for ontological loneliness, 108–09
    contemplative attention and, 57, 59–60, 62–63
    Domanski and, 49–53
    erotics as first philosophy, 61–62
    philosophy and, 56–57
    Plato on, 56, 57–58
    pre-life of as ascesis, 109–12
Poitras, Edward, 66
Poitras, Robin, 66
politics, 3–4, 42, 116–17, 119–20
Porete, Marguerite, 140–41
poverty, 98–99, 100, 101–02
prayer, 57
primary process and primary imagination, 33–35
Proclus, 26–27, 37, 123
pseudo-Dionysius the Areopagite
    about, 30, 132, 145
    on angels (celestial hierarchy), 155, 158
    The Celestial Hierarchy, 153, 154, 158
    on ceremony and hunger for divine, 138–40
    The Ecclesiastical Hierarchy, 138
    on light, 154–55
    Maximus the Confessor and, 143
    on theandric energy, 144
psychology (psychoanalysis), 44, 103–04

reading. See *lectio divina*
reason, 66, 107–08, 114–15, 119, 136–37, 145
reconciliation, 45–46, 145
religion, 56–57, 59, 132–33
representation of place, 13, 29
residential schools, 131–32, 133, 145
Rich, Adrienne, 4
Riel, Louis
    Massinahican, 13–14, 66
Riel, Sara, 14
Robinson, Marilynne, 31
Robinson, Tim, 51
Roman Catholic Church
    Benedictine tradition, 17, 102, 110–11
    Cistercian tradition, 110–11
    esoteric, suppression of, 140–42
    Francis of Assisi (Franciscan tradition), 21, 97–99, 100–01, 102
    residential schools and, 131–32, 133
    See also Christianity
Rose, Gillian
    about, 26, 30
    on civil society, 116–17
    on innocence, 116
    on mourning, 117, 120
    on *Phenomenology of Spirit* (Hegel), 118–19

on philosophy's broken middle, 43,
    114, 115–16, 119
on philosophy's dismantling, 113–14,
    114–15
political model sought by, 119–20
on spirit, 117–18
Royce, Josiah, 46
Rubenstein, Richard, 6–7
Rumi, 40
Rycroft, Charles, 34

sadness, 4–6, 17–18, 45, 93, 110. *See
    also* mourning; *penthos*
Scott, Peter Dale, 59
self
    acedia vs. poverty, 101–02
    'ayan (ultimate identity of a thing)
        and, 84–86
    eros, practice of, and, 82–84
    split self, 42–44
    ungraspability of, 77–79, 80, 81–82,
        89
    visionary recital and, 87–89
    See also contemplation; interiority
SENĆOŦEN language, 1, 2, 132. *See also*
    W̱SÁNEĆ
settler colonialism, 2, 44–45, 76, 132,
    133, 137, 141, 155. *See also*
    imperialism
Shanks, Andrew, 114
Skaay (Haida poet), 57
    "The One They Hand Along," 59
sloth, 94. *See also* acedia
"slow philosophy" movement, 30
SṈ,AḴE (Mount Tolmie), 92, 145,
    153–54
Socrates
    on *atopos*, 62
    Diotima and, 88, 105, 133–35, 140,
        141, 142, 145–46
    on love, 106
    on philosophy, 59
    Plato and, 125

Pythia's revelation and, 38, 126
in *Republic*, 58
sharing the fruits of his
    contemplation, 36–37, 38,
    39–40
in *Symposium*, 104
teaching moments, 28–29, 89
somatic repetition, 110–12
sorrow. *See* mourning; *penthos*; sadness
Spinoza, Baruch, 26
spirit, 43, 117–18, 126
spiritual direction, 12–13, 20
stabilitas, 9, 95, 96, 102
Steinbeck, John
    The Grapes of Wrath, 31
Stó:lō, 10
Sufism, 30, 36, 87. *See also* Ibn 'Arabi
Suhrawardi, 29, 72
Swedenborg, Emmanuel, 72
Szumigalski, Anne
    "Theirs Is the Song," 55–56, 63–64

Taoism, 30, 46, 109
Tao Te Ching, 1, 19, 32, 77
tasting (finding), 68–69, 80–81, 87
ta'wil (interpretation), 71, 72, 76, 78,
    84–86, 88, 137
teaching, 25–27, 28–29
Tecumseh Republic, 15–16
Teilhard de Chardin, Pierre, 141
Teresa of Avila, 40, 43, 126, 139
theandrism, 143–45
theurgy, 12, 13–14, 37, 138–39, 141–42
Thomas Aquinas
    Summa Theologica, 37
Thoreau, Henry David, 44
    Walden, 47
Tillich, Paul, 94
transfixity, 4
Trinity Western University, 2–3

Varnhagen, Rahel, 115, 118
visionary recitals, 27, 29, 78, 87–89

von Tiesenhausen, Peter, 4

Waldrep, G.C., 40
Walker, Michelle Boulous, 30
Ward, Benedicta
    The Sayings of the Desert Fathers, 96
weather, extreme, 131. *See also* climate
    change
Weil, Simone, 12, 15, 26, 30
Western intellectual tradition, 66,
    75–76, 106–08, 153
whiteness, 132, 145
wilderness, of language, 109

Wittgenstein, Ludwig, 26
Wordsworth, William, 44
wren, winter, 158
W̱SÁNEĆ, 13, 17–18, 30, 40, 44, 47,
    91–92. *See also* SENĆOŦEN
    language

Xi Chuan, 34

Zeno, 29, 37, 134
Zohar, 2, 13, 14–15
Zwicky, Jan, 26, 30, 33, 34

Other Titles from University of Alberta Press

The Larger Conversation
Contemplation and Place
TIM LILBURN
Philosophical commentaries on the difficult task of forming a deep, respectful relationship with the land.

Making Wonderful
Ideological Roots of Our Eco-Catastrophe
MARTIN M. TWEEDALE
Documents how the West came to have an ideology that has promoted environmentally destructive economic expansion.

Monitoring Station
SONJA RUTH GRECKOL
Feminist experimental lyric poetry that embodies the passage of a damaged world across generations.
Robert Kroetsch Series

More information at uap.ualberta.ca

www.ingramcontent.com/pod-product-compliance
Ingram Content Group UK Ltd.
Pitfield, Milton Keynes, MK11 3LW, UK
UKHW011605021025
463508UK00002B/156